SpringerBriefs in Earth System Sciences

Series editors

Gerrit Lohmann, Bremen, Germany
Lawrence A. Mysak, Montreal, Canada
Justus Notholt, Bremen, Germany
Jorge Rabassa, Ushuaia, Argentina
Vikram Unnithan, Bremen, Germany

More information about this series at http://www.springer.com/series/10032

Nathan Paldor

Shallow Water Waves on the Rotating Earth

 Springer

Nathan Paldor
Fredy and Nadine Herrmann Institute
  of Earth Sciences
The Hebrew University of Jerusalem
Jerusalem
Israel

ISSN 2191-589X   ISSN 2191-5903   (electronic)
SpringerBriefs in Earth System Sciences
ISBN 978-3-319-20260-0   ISBN 978-3-319-20261-7   (eBook)
DOI 10.1007/978-3-319-20261-7

Library of Congress Control Number: 2015941492

Springer Cham Heidelberg New York Dordrecht London

Printed on acid-free paper

Springer International Publishing AG Switzerland is part of Springer Science+Business Media
(www.springer.com)

Preface

This book describes new theoretical advances made in recent years that yielded new analytical solutions of the rotating shallow-water equations. The intended readership of the book consists of graduate students and scientists in the fields of geophysical fluid dynamics, physical oceanography, dynamical meteorology, and applied mathematics who study or employ the shallow-water equations. The new dispersion relations and meridional amplitude structures of the waves derived in this book can be applied to observations in the atmosphere and ocean and they also provide alternatives to the spherical harmonics basis of global-scale spectral numerical models. The book originates from years of teaching courses in dynamical meteorology and physical oceanography at the Hebrew University of Jerusalem as well as at the University of Miami and Johns Hopkins University during my sabbatical stays there. In textbooks of Physical Oceanography and Dynamical Meteorology the theory of Kelvin, Inertia-Gravity (Poincaré) and Planetary (Rossby) Waves are developed starting from a different set of equations for each wave type instead of a single set of equations in which a certain parameter is set to a specific value or limiting value (e.g., zero, one, infinity). An additional issue with the traditional approach is that equatorial waves cannot be obtained from the mid-latitude waves by setting to zero the central latitude in the solutions for mid-latitude waves. This can be readily verified by realizing that the harmonic meridional structure of mid-latitude waves' amplitudes does not reduce to the Hermite functions structure of equator waves when the central latitude of the mid-latitude theory is set to zero.

In this book, I attempt to bridge the gap between equatorial wave theory and mid-latitude wave theory by developing a Schrödinger eigenvalue equation whose eigenvalues yield the phase speeds and the eigenfunctions yield the amplitude structure of both Inertia-Gravity and Planetary waves. This Schrödinger equation formulation was originally applied in 1966 by Taro Matsuno in his development of wave theory on the equatorial β-plane. The same approach is extended to mid-latitude wave theory which is traditionally developed on the f-plane for all waves except Rossby waves for which the theory is developed on the mid-latitude β-plane. In the unified approach, all waves are developed on the latter plain. In this approach, Kelvin waves are a particular wave type that solves a degenerate case where the second-order eigenvalue problem degenerates to a first-order equation that determines the amplitude meridional structure, while the dispersion relation is determined from a wave equation in the zonal direction that is identical to the wave equation for gravity waves of the shallow water equations in one-dimension without rotation.

In contrast to Cartesian coordinates where the Schrödinger eigenvalue equation approach to shallow water waves offers an alternate (perhaps more elegant, general, and accurate) derivation to the traditional harmonic derivation, the application of this approach in spherical coordinates yields analytical results that cannot be derived otherwise. The first to recognize the importance of formulating an eigenvalue equation for the rotating shallow-water equations over a sphere was Michael S. Longuet-Higgins who successfully derived such equations in his seminal 1968 paper on "Laplace Tidal Equations Over a Sphere." His detailed analysis and extensive numerical calculations (which is amazing considering the meager computing power that was available to him in the 1960s) of the exact equation make it clear that no progress can be made on the subject without first simplifying the eigenvalue equations. The necessary simplification was recently achieved for parameter values typical to earth by maintaining terms in the spherical problem that have counterparts on a plane. The accuracy of the solutions of the resulting simplified eigenvalue equation on a sphere provides the justification for the approximation more than the formal asymptotic expansion.

My research associate, Dr. Andrey Sigalov, along with my graduate students at the Hebrew University of Jerusalem, Shira Rubin, Yair De-Leon, and Ofer Shamir, have contributed greatly to the development of the ideas presented here and to the analysis and numerical calculations that are integral elements of the theory and its presentation to the readers of this book. I am indebted to them for their help in making the theoretical advances and consider them my partners in this endeavor. Though they share with me the major theoretical advancement, I am solely responsible for errors that have not been uprooted from this monograph. My colleague and close friend Prof. Hezi Gildor of the Hebrew University of Jerusalem encouraged me to write this book. I am grateful to him for the impetus he provided to me when the idea of writing this book was first conceived and for his encouragement during the time when this book was written. The Bogen Family Foundation supported my sabbatical stay at University of Pennsylvania during the spring semester of 2015 during which this book was completed.

Jerusalem
June 2015

Contents

Chapter 1
Introduction

This chapter reviews the state of the shallow water wave theory as it appears in textbooks on the subject, which was developed in recent decades. In this traditional theory, the meridional structure of the amplitude of the zonally propagating waves is described only by harmonic functions (sin, cos, or exponentials) in mid-latitudes. This mid-latitude amplitude structure differs qualitatively from that on the equatorial β-plane where the structure is described by Hermite functions. In sharp contrast to these two planar theories, no explicit/analytic expressions exist for zonally propagating waves on a sphere and the meridional amplitude structure can only be calculated numerically for known values of both the zonal wave number and the phase speed (or frequency). The governing equations employed in the derivations of the waves' dispersion relation and meridional amplitude structures employ somewhat different approximations that vary with the geometry under study (e.g., a mid-latitude f-plane, a β-plane, or a sphere) and with the type of wave and in order to obtain a coherent scenario for all of these geometries and wave types, it is useful to start with a review of the existing theories.

The vectorial form of the linearized shallow water equations (SWE) in the presence of rotation (that adds the Coriolis term to the acceleration in the momentum equations) is:

$$\frac{\partial \mathbf{V}}{\partial t} + f\,\hat{k} \times \mathbf{V} = -g\nabla\eta$$
$$\frac{\partial \eta}{\partial t} = -H\nabla\cdot\mathbf{V}, \tag{1.1}$$

where V is the horizontal velocity vector (i.e., the velocity along a plane perpendicular to the unit vector $\hat{k}$ oriented parallel to the radial/vertical direction) whose components in the zonal (λ, longitude, East) and meridional (ϕ, latitude, North) directions are u and v, respectively; f is the latitude-dependent Coriolis frequency ($= 2\Omega \sin\phi$, where Ω is earth's rotation frequency); H is the constant mean thickness of the fluid; η is the deviation of total height, h, from H (i.e., $h = H + \eta$ is the actual upper surface of the fluid, which varies in space and time); and g is the gravitational constant (or the reduced gravity in an equivalent barotropic, two-layer fluid). The set (1.1) is also known as Laplace tidal equation (LTE) as it describes the

© The Author(s) 2015
N. Paldor, *Shallow Water Waves on the Rotating Earth*,
SpringerBriefs in Earth System Sciences, DOI 10.1007/978-3-319-20261-7_1

horizontal dynamics in response to tidal forcing (e.g., the gravitational potential in the ocean or the thermal forcing in the atmosphere) that affects directly the dynamics in the vertical (radial in the case of a sphere) direction only.

Equation (1.1) provides the very basic description of small amplitude fluid dynamics in the presence of rotation. The first (vectorial) equation is the application of Newton's second law of motion to a unit volume of fluid of constant (potential) density in a rotating frame of reference (where the frequency of rotation is Ω) subject to the pressure gradient force when the pressure is in hydrostatic balance. The second equation guarantees the conservation of mass when the fluid is assumed to be incompressible. This set is nothing but Euler's equations for a two-dimensional flow where the linearized acceleration in the rotating frame includes the Coriolis term that accounts for the transformation to an inertial frame where Newton's second law of motion should be applied.

Wave solutions of (1.1) were found in a variety of coordinate systems, physical setups and geometries. Most of the theoretical advances in this field were made in Cartesian coordinates that are relevant to a tangential plane to earth's ellipsoidal surface at some latitude ϕ_0 (Fig. 1.1). This is a natural outcome given the relative simplicity afforded by this coordinate system compared to the much more complex spherical coordinates. In mid-latitudes, the tangential plane in which the Coriolis frequency, f, is taken to be a constant, f_0, is called the f-plane; when f is assumed to vary linearly with the northward coordinate, y, i.e., when $f = f_0 + \beta y$ (where $\beta = \partial f / \partial y$ is taken to be a constant) the same plane is called the β-plane. Wave theories were also developed on the unbounded equatorial β-plane (where f_0 is set equal to 0), on the whole sphere, and on a general two-dimensional curved surface of

Fig. 1.1 The tangential plane to earth at the central latitude ϕ and the Cartesian coordinate system (East, North) = (x, y) in this plane. The perpendicular (*vertical*) coordinate, z, is aligned parallel to earth's radius that passes through the tangential point with positive values directed away from earth's center

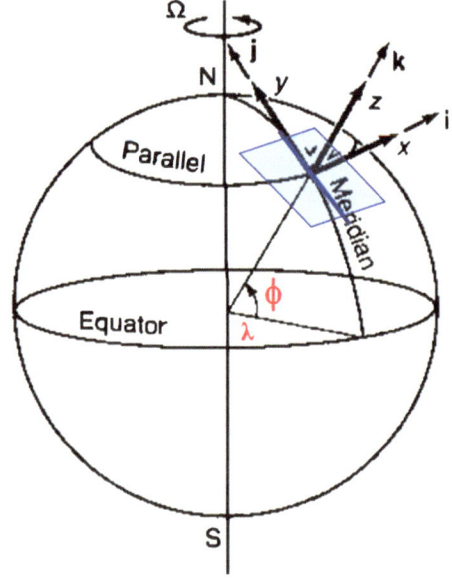

unspecified geometry (i.e., an invariant theory). In the next subsections, I summarize the classical (and in cases overly simplistic) results obtained in various setups in these coordinate systems.

1.1   Cartesian Coordinate

In Cartesian coordinates (x, y) where x points eastwards (and the velocity component in this direction is u) and y points northwards (and the velocity component in this direction is v), system (1.1) takes the form:

$$
\begin{aligned}
\frac{\partial u}{\partial t} - (f_0 + \beta y)v &= -g\frac{\partial \eta}{\partial x}, \\
\frac{\partial v}{\partial t} + (f_0 + \beta y)u &= -g\frac{\partial \eta}{\partial y}, \\
\frac{\partial \eta}{\partial t} &= -H\left(\frac{\partial u}{\partial x} + \frac{\partial v}{\partial y}\right).
\end{aligned}
\tag{1.2}
$$

In this set, $f_0 = 2\Omega \sin \phi_0$ and $\beta = \partial f/\partial y = 2\Omega \cos \phi_0/a$ where ϕ_0 is the central latitude at which the plane is tangential to earth and a is earth's radius (see Fig. 1.1).

1.1.1   Waves in a Mid-latitude Channel

Three wave types are known to exist in a mid-latitude zonal channel of width L, i.e., when the meridional domain extends from $y = -L/2$ to $y = L/2$ and the northward velocity component, v, is required to vanish at $y = \pm L/2$ on the β-plane. The wave types are derived by letting the solution for u, v, and η in (1.2) vary in the form $\mathrm{Re}\{Ae^{i(kx+ny-\omega t)}\}$ where A is a (possibly complex) constant amplitude and where the meridional wave number n is quantized so as to guarantee that v vanishes at $y = \pm L/2$ (e.g., when A is real, $n = (2N + 1)\pi/L$ where N is an integer number: $N = 0, 1, 2, 3, \ldots$). The designation $\mathrm{Re}\{\}$ is omitted when it is trivially implied by the context, and when the relative phases of different functions have to be noted the real sin and cosine functions will be used instead of the complex exponential form. The dispersion relation $\omega(k, n)$ and the relations between the amplitudes of u, v and η provide the defining characteristics of the various wave types. For a detailed derivation of the mid-latitude waves, the reader is referred to textbooks, e.g., Vallis (2006), Pedlosky (1987) and Cushman-Roisin (1994).

The first wave type is Kelvin waves that are derived by letting $v = 0$ everywhere in (1.2) so the boundary conditions at $y = \pm L/2$ are trivially satisfied for any value of the meridional wave number, n, i.e., n is not quantized and can be complex.

Setting $v = 0$ in the first and last equations of (1.2) and eliminating one of them yield the dispersion relation of gravity waves of a non-rotating system:

$$\omega = \pm\sqrt{gHk} \iff C_g \equiv \frac{\omega}{k} = \pm\sqrt{gH}, \tag{1.3}$$

where C_g is the (gravity wave) phase speed in the x-direction. Specifically, cross-differentiating the first and last equations to eliminate either u or η yields the second-order wave equation $\theta_{tt} = (gH)\theta_{xx}$ where θ is either η or u so the (x, t) variation of u and η is given by $e^{ik(x \pm C_g t)}$, while the y-dependence of their amplitudes is identical since for $v = 0$ the first equation in (1.2) yields $C_g u = g\eta$. The second equation in (1.2) then determines the y-dependence of the amplitudes of u and η which is given by: $e^{-\frac{1}{C_g}\int (f_0 + \beta y)dy} = e^{-\frac{1}{C_g}\left(f_0 y + \frac{\beta y^2}{2}\right)}$. With $C_g = \pm(gH)^{1/2}$, the zonal velocity and height fields of these waves are given by (η_0 is an arbitrary amplitude of $\eta(x, y, t)$):

$$v = 0,$$

$$u = \frac{g}{C_g}\eta_0 e^{\frac{-1}{C_g}\left(f_0 y + \frac{\beta y^2}{2}\right)} e^{ik(x - C_g t)}, \tag{1.4}$$

$$\eta = \eta_0 e^{\frac{-1}{C_g}\left(f_0 y + \frac{\beta y^2}{2}\right)} e^{ik(x - C_g t)}.$$

The two signs of C_g define two waves: the amplitude of the wave with $C_g = +(gH)^{1/2} > 0$ decays when y increases so its amplitude is maximal at $y = -L/2$, while the amplitude of the wave with $C_g = -(gH)^{1/2} < 0$ decays with the decrease in y so its amplitude is maximal at $y = +L/2$.

The second type of waves is Inertia-Gravity waves (also known as Poincaré waves) and these waves are derived from (1.2) by setting $(u, v, \eta) = (u_0, v_0, \eta_0) \operatorname{Re}\left\{e^{i(kx + ny - \omega t)}\right\}$ where the (high) frequency satisfies $\omega \geq f_0 \gg \beta L/2$ so the local acceleration is larger than the nearly constant, Coriolis frequency. Substituting these expressions in (1.2) and requiring that the resulting linear equations for the amplitudes u_0, v_0 and η_0 have a nonzero solution then yields the dispersion relation of these waves:

$$\omega = \pm\sqrt{gH(k^2 + n^2) + f_0^2}. \tag{1.5}$$

The frequency in (1.5) is the Pythagorean sum of the inertial frequency f_0 and the gravitational frequency $(gH)^{1/2}K$ given in (1.3) where $K = (k^2 + n^2)^{1/2}$ is the total wave number, so for large k, (1.5) approaches (1.3). The u, v and η fields are given by:

$$u = \frac{v_0}{\omega^2 - gHk^2}\left(\omega f_0 \sin\left(\frac{N\pi}{L}\left(y+\frac{L}{2}\right)\right) - nkgH\cos\left(\frac{N\pi}{L}\left(y+\frac{L}{2}\right)\right)\right)\cos(kx-\omega t),$$

$$v = v_0 \sin\left(\frac{N\pi}{L}\left(y+\frac{L}{2}\right)\right)\sin(kx-\omega t),$$

$$\eta = \frac{v_0}{\omega^2 - gHk^2}\left(kf_0 H \sin\left(\frac{N\pi}{L}\left(y+\frac{L}{2}\right)\right) + \omega nH\cos\left(\frac{N\pi}{L}\left(y+\frac{L}{2}\right)\right)\right)\cos(kx-\omega t),$$

$$(1.6)$$

where $N\pi/L, N = 1,2,3\ldots$ is the meridional wave number, n and v_0 is an arbitrary normalization constant. Note that the denominator of the expressions for u and η i.e., $\omega^2 - gHk^2$ never vanishes since according to (1.5) it equals the definite positive combination $gHn^2 + f_0^2$.

The third wave type is Planetary waves (also known as Rossby waves) that are also obtained from (1.2) by setting $(u, v, \eta) = (u_0, v_0, \eta_0)\, e^{i(kx+ny-\omega t)}$, but with low frequency, $\omega \ll f_0$, so the latitude-dependent part of f, i.e., βy (which is bounded by $\beta L/2$), cannot be neglected. To first order in β (assumed to be a small parameter), the dispersion relation of these waves is:

$$\omega = \frac{-\beta k}{k^2 + n^2 + \frac{f_0^2}{gH}}, \qquad (1.7)$$

which vanishes for $\beta = 0$, in which case it degenerates to the steady geostrophic balance on the f-plane. To zeroth order in β, the u, v and η fields are:

$$u = -\frac{N\pi}{L}\frac{1}{k}v_0\cos\left(\frac{N\pi}{L}\left(y+\frac{L}{2}\right)\right)\cos(kx-\omega t),$$

$$v = v_0 \sin\left(\frac{N\pi}{L}\left(y+\frac{L}{2}\right)\right)\sin(kx-\omega t), \qquad (1.8)$$

$$\eta = -\frac{f_0}{gk}v_0\sin\left(\frac{N\pi}{L}\left(y+\frac{L}{2}\right)\right)\cos(kx-\omega t),$$

where $N\pi/L, N = 1,2,3,\ldots$ is the meridional wave number, n and v_0 is an arbitrary normalization constant. In this solution, both v and η vanish at $y = \pm L/2$ though only v is required to do so.

Figure 1.2 shows the dispersion relation of the three wave types: Top panel: Non-dispersive Kelvin waves, (1.3), and Inertia-Gravity waves, (1.5); Bottom panel: low-frequency Planetary waves, (1.7). The scales of the ordinates in the two panels are drastically different.

Three remarks should be made with regard to the above expressions by the classical mid-latitude theory.

1. Both Kelvin waves and Inertia-Gravity waves are traditionally derived from (1.2) by setting $\beta = 0$, i.e., on the f-plane. While the extension of Kelvin waves

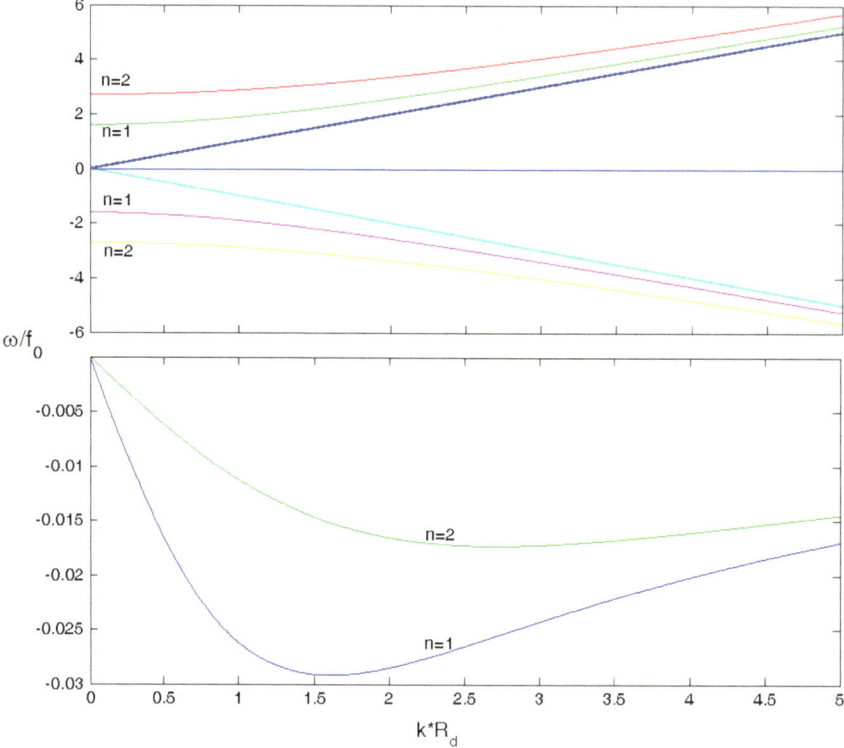

Fig. 1.2 The dispersion relations ω/f_0 as a function of kR_d (for $R_d = (gH)^{1/2}/f_0 = 600$ km) for the indicated meridional mode numbers, n, for Kelvin and Poincare waves (*upper panel*) and Rossby waves (*lower panel*) in a 1500 km wide channel centered on $45°$

    to the β-plane in (1.4) is fairly simple, this is not all the case for Inertia-Gravity waves and it is unclear how the dispersion relation of these waves and the y-structure of their amplitude will be altered for $\beta > 0$.

2. The expressions for the dispersion relation and amplitude structure of Planetary waves contain only the two constant f_0 and β, while the original set of equations, (1.2), contains a y-dependent coefficient, $f(y)$, that varies linearly with y. While in general, a linear function is fully describable by two constants, the solutions of a differential system with non-constant coefficients can never be harmonic functions (the reader should consider $\sin(ny)$ and $\cos(ny)$) since these functions solve only differential equations that have constant coefficient.

3. As mentioned above, the solution (1.8) for Planetary waves is accurate only to zeroth order in β and setting $\beta = 0$ in the dispersion relation of these waves, (1.7), shows that (1.8) degenerates to the steady geostrophic solution on the f-plane for $\beta = 0 = \omega$. Corrections to (1.8) were calculated in Heifetz et al. (2007) by decomposing the solution into the geostrophic part, (1.8) with $\omega = 0$,

and an ageostrophic part which is proportional to β. By solving higher order terms in β encountered when (1.8) is substituted in (1.2), Heifetz et al. (2007) showed that for small values of $b = \frac{1}{2}\beta L/f_0$ (i.e., when the maximal relative change in f across the channel is small), the correction terms have the form of low-order polynomials in y times $\sin(ny)$ and $\cos(ny)$.

1.1.2 Waves on the Equatorial β-Plane

The equatorial case is obtained from (1.2) by setting $\phi_0 = 0$, i.e., $f_0 = 0$ and $\beta = 2\Omega/a$. A complete theoretical study of case of an infinite (i.e., unbounded) β-plane is given in Matsuno (1966) and no modifications to this seminal and thorough contribution were brought up since its publication nearly 50 years ago. At about the same time, Lindzen (1967) rederived Matsuno's results on Planetary and Inertia-Gravity waves in the forced problem in which a more realistic form of the vertical variation is included, while Holton and Lindzen (1968) added the special Kelvin wave solution to that same problem by setting $v = 0$. With proper scaling, the eigenvalue equation that determines the frequency and meridional variation of the amplitudes of the zonally propagating wave solutions is precisely that of Harmonic Oscillator of Quantum Mechanics. The immediate implication of this elegant formulation is that the amplitude eigenfunctions are Hermite functions (i.e., Hermite polynomial multiplied by a Gaussian) and the eigenvalues equal $2n + 1$, where $n = 0, 1, 2, \ldots$ is the meridional mode number. These eigenvalues determine the phase speeds of Planetary and Inertia-Gravity waves via the roots of a cubic polynomial whose coefficients include the zonal wave number, k, and $C_g = (gH)^{1/2}$. Kelvin wave with the same dispersion relation as in (1.3) can be formally obtained from this dispersion relation by setting $n = -1$ but can be also developed directly from the original equations by setting $v(y) = 0$.

The $n = 0$ case of a westward-propagating wave of Matsuno's theory yields a special wave type that has no counterpart in the mid-latitude theory—the mixed mode. On the dispersion diagram shown in Fig. 1.3, this mode appears as an Inertia-Gravity wave at small k (intersecting the frequency ordinate at a finite value) and as a Planetary wave at large k (approaching the abscissa as k tends to infinity). The nature of the transition from an Inertia-Gravity wave to a Planetary wave at the wave number where the wave's phase speed equals $-C_g$ is not quite clear.

Channel theories have also been developed on the equatorial β-plane, but these were primarily numerical (Cane and Sarachik 1976, 1977, 1979) or combined numerical/analytical (Erlick et al. 2007). The latter study has demonstrated that the mixed mode consists of two separate modes even in a channel that extends to 1.4 rad (channel half-width of $1.4a$), i.e., far beyond the range of applicability of the equatorial β-plane approximation.

Two fundamental issues need to be pointed out. The first point arises when one compares the results of the equatorial β-plane with those in mid-latitudes. Even

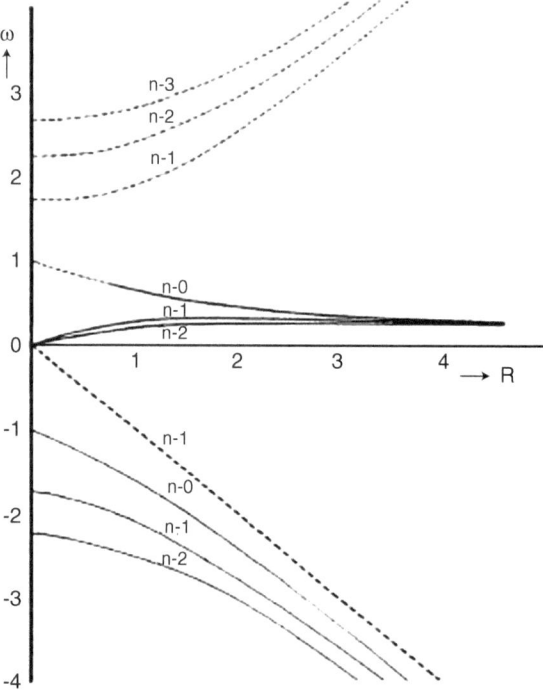

Fig. 1.3 The dispersion relation of zonally propagating waves on the equatorial β-plane, $e^{i(kx+wt)}$, i.e., eastward-/westward-propagating waves have negative/positive frequencies, respectively (from Matsuno 1966). *Thin solid lines* eastward-propagating Inertia-Gravity waves. *Thin dashed lines* westward-propagating Inertia-Gravity waves. *Thick solid lines* Planetary (Rossby) waves. *Thick dashed line* ($n = -1$) Kelvin wave. The westward-propagating $n = 0$ wave (positive frequency) changes from *thin dashed* (Inertia-Gravity) to *thick solid* (Planetary) lines at the intersection point with a reflection of the $n = -1$ line (Kelvin wave). Permission from Meteorological Society of Japan: J. Met. Soc. Japan, 1966, 44(1), 25–43

though the limit $\phi_0 = 0$ (i.e., $f_0 = 0$) is a regular limit of (1.2), this is not the case in the solutions since by setting $f_0 = 0$ in the amplitude structures (1.6) and (1.8), one does not obtain the Hermite functions of the equatorial theory (in fact, the Harmonic Inertia-Gravity waves in mid-latitudes are hardly modified by setting $f_0 = 0$). This dissonance between the oscillatory solutions in mid-latitudes and fast decaying Hermite functions on the equator is unrelated to the quantization of the meridional wave number, n, in the mid-latitude channel theory—it only reflects the overly simplistic manner in which the variation of $f(y)$ is treated in the mid-latitude theory where $f(y)$ is taken to be a constant ($= f_0$) even though its derivative (i.e., β) does not vanish.

The second issue has to do with the relevance of the mixed mode of the infinite equatorial β-plane to a channel and to a sphere. Since this mode remains separated (i.e., there is no mixed mode) in a channel with walls located close to the poles, one wonders whether this mode is relevant at all to a sphere. An alternative view is that it is merely an artifact of the planar approximation to earth's curved shape in which $\sin(\phi)$ of the Coriolis frequency is expanded to first order, while the $\tan(\phi)$ term of the divergence operator in the continuity equation (the second equation in (1.1), see also the third equation in (1.9) below) is kept at zeroth order only in ϕ, i.e., the linear y-term that arises from the expansion of $\tan(\phi)$ is neglected altogether.

1.2 The Eigenvalue Equation for Waves on a Rotating Sphere

In spherical coordinate, the set (1.1) is known as LTE and it takes the form:

$$\frac{\partial u}{\partial t} - 2\Omega v \sin \phi = -\frac{g}{a \cos \phi} \frac{\partial \eta}{\partial \lambda},$$

$$\frac{\partial v}{\partial t} + 2\Omega u \sin \phi = -\frac{g}{a} \frac{\partial \eta}{\partial \phi}, \qquad (1.9)$$

$$\frac{\partial \eta}{\partial t} = -\frac{H}{a \cos \phi} \left(\frac{\partial u}{\partial \lambda} + \frac{\partial (v \cos \phi)}{\partial \phi} \right).$$

In the seminal study of Longuet-Higgins (1968), zonally propagating wave solutions of this set, $e^{i(k\lambda - \omega t)}$, were transformed into various eigenvalue equations by eliminating two of the three variables u, v and η. Since the first equation in the set (i.e., the equation for the time evolution of u) does not contain derivatives with respect to ϕ, the assumption of zonally propagating wave solutions results in a 2nd-order differential equation in ϕ for the meridional structure of the waves' amplitude. By eliminating of u and v from (1.9), Longuet-Higgins (1968) derived the following eigenvalue equation for η:

$$\Xi(\eta) = \gamma \eta \qquad (1.10)$$

where the linear second-order differential operator, Ξ, is given by:

$$\Xi = \frac{2\Omega}{\omega \cos \phi} \left\{ \frac{\partial}{\partial \phi} \left[\frac{4\Omega^2}{4\Omega^2 \sin^2 \phi - \omega^2} \left(k \sin \phi + \frac{\omega}{2\Omega} \cos \phi \frac{\partial}{\partial \phi} \right) \right] \right.$$
$$\left. + \frac{4\Omega^2 k}{\omega^2 - 4\Omega^2 \sin^2 \phi} \left[\frac{\omega k}{2\Omega \cos \phi} + \sin \phi \frac{\partial}{\partial \phi} \right] \right\}$$

and where the eigenvalue, γ, is defined by: $\gamma = \frac{4\Omega^2 a^2}{gH}$. This parameter is also known as Lamb number or Lamb parameter.

This formidable eigenvalue equation is too complex to be solved analytically, but even if an eigensolution of this equation is somehow found and both the eigenvalue $\gamma(k, \omega)$ and the corresponding eigenfunction $\eta(\phi; k, \omega)$ (i.e., the function $\eta(\phi)$ with given k and ω) are known for any values of k and ω it is not clear how to invert the $\gamma(k, \omega)$ relation to obtain the dispersion relation $\omega(k; \gamma)$. Solutions of the eigenvalue equation (1.10) were calculated numerically in Longuet-Higgins (1968) and subsequently in Moura (1976), but with the exception of some particular asymptotic cases (e.g., non-rotating or near the equator) these numerical calculations have not yielded explicit expressions for the dispersion relations and meridional structures of these waves.

1.3 Invariant Theory and Its Application to a Sphere and a Spheroid

Invariant wave theories of the linear shallow water equations (LSWE) involve more abstract tools that need to be adapted from the field of differential geometry and applied to geophysical fluid dynamics. The geometry-independent wave theories derived from these adaptations can then be applied to complex, but relevant, geometries such as a sphere or a spheroid (i.e., an ellipsoid of revolution) for certain types of waves. Müller and O'Brien (1995) were the first to employ a differential geometry formulation of the LSWE and this formulation underscored the essential role played by the prolate spheroidal angular wave functions (see Abramowitz and Stegun 1972) in wave theory of LSWE on a sphere. However, the theory developed in Müller and O'Brien (1995) failed to provide explicit expressions for the dispersion relation and the spatial structure of either Inertia-Gravity waves or Planetary waves. Paldor and Sigalov (2011) formulated the LSWE on a smooth 2D Riemannian surface of arbitrary shape with a general Coriolis parameter oriented locally perpendicular to that surface. Though this is also implicit in the vector formulation of the LSWE, e.g., (1.1), the tools available from 3D vector analysis are not helpful in the investigation of 2D problems on curved surfaces. The 2D formulation enabled the application of tensor calculus that yielded approximate expressions for the dispersion relation and the meridional amplitude structure for both Inertia-Gravity and Planetary waves where the distinction between the two wave types is based on the magnitude of their frequencies. The general approximate formulae obtained from the application of these ideas have yielded explicit expressions on a plane and on a sphere where on a plane the resulting expressions have degenerated to the known expressions while on a sphere they degenerate to the particular expressions derived by Longuet-Higgins (1968). The same general expressions derived by Paldor and Sigalov (2011) were subsequently applied by Paldor and Sigalov (2012) to spheroids and this application demonstrated that the errors introduced by approximating earth's spheroid shape by a sphere are bounded by earth's eccentricity. The implication of this estimate is that the 0.3 % error associated with earth's eccentricity is too small to be of importance in the applications of the spherical wave theory to the rotating earth.

1.4   A Unified Approach to Wave Theory of the Shallow Water Equations

With the exception of the invariant approach described in Sect. 1.3 in each of the setups described above in Sects. 1.1 and 1.2, the theory was developed using a different eigenvalue formulation: Harmonic Oscillator of Quantum Mechanics on the equator, a (nearly) constant coefficient equation in mid-latitudes and a highly complex and unsolvable equation on the entire sphere. This is surprising in view of the fact that the basic set, (1.1), is identical in all cases as is the form assumed for the solution (zonally propagating wave) which makes one wonder whether a unified approach that yields a similar, if not identical, eigenvalue equation in all the cases outlined above can be developed.

While the unsolvable eigenvalue equation of waves on the entire sphere is unique to the SWE in spherical coordinates the eigenvalue equation in a mid-latitude channel is a genuine constant coefficient equation on the f-plane where Inertia-Gravity and Kelvin waves exist. Even in the case of Planetary waves whose frequency is linear with β, the solution behaves essentially as a solution of constant coefficient equation in which two of the constants are f_0 that replaces $f(y)$ everywhere and $\beta(=\partial f/\partial y)$ which is assumed to be nonzero even though f is set to a constant elsewhere. It is unlikely that any one of these approaches can be applied to construct a general paradigm for waves on the rotating earth in all geometries: The first is too complex to yield explicit solutions and the latter is too simple to provide solutions other than purely oscillatory ones. Only the approach that leads to the eigenvalue equation of equatorial waves, developed in Matsuno (1966) might be generalized to yield wave solutions in other geometries (spherical) and setups (bounded domains such as channels). This approach has also yielded exact solutions of a second-order eigenvalue equation whose eigenvalues yielded expressions for all three types of waves: Planetary, Inertia-Gravity and Kelvin though Kelvin waves were derived as a singular case of the equation (i.e., the unphysical $n = -1$ case).

In order to generalize to other problems the approach employed by Matsuno (1966) in his construction of the eigenvalue equation for waves on the infinite equatorial β-plane, it is helpful to highlight the essential elements of this approach.

1. The eigenvalue equation is obtained by letting the solutions vary as zonally propagating waves (i.e., $e^{ik(x-Ct)}$ on the equatorial β-plane) which yields a second-order eigenvalue equation in y.
2. In the planar equatorial problem, the eigenvalue equation has the form of a time-independent Schrödinger equation in which the potential is given by the square of the latitude-dependent Coriolis frequency.
3. The energy of the Schrödinger equation, E_n, yields expressions for the frequencies of three wave types via the roots of a cubic $C(E_n)$ relation. The three families are as follows: eastward- and westward-propagating Inertia-Gravity waves and westward-propagating Planetary waves.

In the rest of this book the Schrödinger equation approach developed by Matsuno (1966) on the infinite equatorial β-plane is generalized to a channels on the mid-latitude and equatorial β-plane as well as to the entire sphere and to channels on a sphere. In addition, the case of an infinite equatorial β-plane is re-examined as the limit of an equatorial channel when the channel width becomes large. In addition to providing explicit solutions on a sphere this new approach also yields phase speeds and meridional structures other than those described above in planar problems and for all wave types, including Harmonic equatorial waves (in addition to the previously found Hermite waves) and Trapped (i.e., not Harmonic) waves in mid-latitudes.

References

Abramowitz M, Stegun IA (1972) Handbook of mathematical functions. Dover Publications, Inc., NY, 1043 p

Cane MA, Sarachik ED (1976) Forced baroclinic ocean motions. I: linear equatorial unbounded case. J Mar Res 34:629–665

Cane MA, Sarachik ED (1977) Forced baroclinic ocean motions. II: linear equatorial bounded case. J Mar Res 35:395–432

Cane MA, Sarachik ED (1979) Forced baroclinic ocean motions. III: the linear equatorial basin case. J Mar Res 37:355–398

Cushman-Roisin B (1994) Introduction to geophysical fluid dynamics. Prentice Hall, NJ, 320 p

Erlick C, Paldor N, Ziv B (2007) Linear waves in a symmetric equatorial channel. Q J Roy Meteorol Soc 133(624):571–577

Heifetz E, Paldor N, Oreg Y, Stern A, Merksamer I (2007) Higher-order corrections for Rossby waves in a zonal channel on the β-plane. Q J Roy Meteorol Soc 133:1893–1898

Holton JR, Lindzen RD (1968) A note on "Kelvin" waves in the atmosphere. Mon Weather Rev 96:385–386

Lindzen RD (1967) Planetary waves on the β-planes. Mon Weather Rev 95:441–451

Longuet-Higgins MS (1968) The eigenfunctions of Laplace's tidal equations over a sphere. Phil Trans Roy Soc Lond A262:511–607

Matsuno T (1966) Quasi-geostrophic motion in the equatorial area. J Meteorol Soc Jpn 44:25–43

Moura AD (1976) The eigensolutions of the linearized balance equations over a sphere. J Atmos Sci 33(6):877–907

Müller D, O'Brien JJ (1995) Shallow water waves on the rotating sphere. Phys Rev E 51:4418–4431

Paldor N, Sigalov A (2011) An invariant theory of the linearized shallow water equations with rotation and its application to a sphere and a plane. Dyn Atmos Oceans 51:28–44

Paldor N, Sigalov A (2012) Linear waves on the spheroidal earth. Dyn Atmos Oceans 57:17–26

Pedlosky J (1987) Geophysical Fluid Dynamics. Springer, Berlin, 710 p

Vallis GK (2006) Atmospheric and oceanic fluid dynamics. Cambridge University Press, Cambridge, 773 p

Chapter 2
Waves in a Channel on the Mid-latitude β-Plane

In this chapter, Trapped wave solutions of system (1.2) subject to the boundary conditions $v(y = \pm L/2) = 0$ will be derived. The amplitudes of these waves decay with distance from the channel wall located closer to the equator (i.e., $y = -L/2$ in the Northern Hemisphere and $y = +L/2$ in the Southern Hemisphere) which will be referred to as the equatorward wall. These Trapped wave solutions supplement the Harmonic waves of Sect. 1.1.1 whose amplitudes spread over the entire channel. Trapped waves dominate the solution when the channel width L is much larger than the radius of deformation $R_d = (gH)^{1/2}/(2\Omega \sin \phi_0)$.

Before attempting to solve the set (1.2), it is instructive to non-dimensionalize it in order to reduce the number of parameters that determine the solution. The scaling procedure is not unique and one simple way of non-dimensionalizing (1.2) is to scale time, t, on $(2\Omega)^{-1}$; x and y on a, earth's radius; u and v on $2\Omega a$; and η on H, the mean layer's thickness. With this scaling $\sin \phi_0$ and $\cos \phi_0$ designate f_0 and β, respectively, while the non-dimensional y is the latitude angle measured relative to ϕ_0 (since the latitude at a dimensional distance y^{dim} is $\phi_0 + y^{\text{nondim}} = \phi_0 + y^{\text{dim}}/a$). These scales transform the set (1.2) to the non-dimensional form (where variables are designated by the same symbols as their dimensional counterparts):

$$\frac{\partial u}{\partial t} - (\sin \phi_0 + y \cos \phi_0)v = -\alpha \frac{\partial \eta}{\partial x},$$

$$\frac{\partial v}{\partial t} + (\sin \phi_0 + y \cos \phi_0)u = -\alpha \frac{\partial \eta}{\partial y}, \qquad (2.1)$$

$$\frac{\partial \eta}{\partial t} = -\left(\frac{\partial u}{\partial x} + \frac{\partial v}{\partial y}\right),$$

where $\alpha = gH/(2\Omega a)^2$ is the only parameter of this non-dimensional differential system that plays the role of g in the dimensional system (1.2). This number is the square of non-dimensional speed of Gravity waves and its inverse is known as Lamb number or Lamb parameter. On earth, $2\Omega a = 931$ m/s and since the speed of gravity waves, $(gH)^{1/2}$, varies from 2 to 3 m/s in a baroclinic ocean to about 200 m/s in a barotropic ocean/atmosphere, the value of α varies between 5×10^{-6}

© The Author(s) 2015
N. Paldor, *Shallow Water Waves on the Rotating Earth*,
SpringerBriefs in Earth System Sciences, DOI 10.1007/978-3-319-20261-7_2

and 5×10^{-2}. The boundary conditions imposed on the solutions of (2.1) are $v(y = \pm \delta\phi/2) = 0$ where $\delta\phi = L/a$ is the non-dimensional channel width expressed as a meridional angle (i.e. in radians).

Since there are no x- or t-dependent coefficients in system (2.1) Fourier theorem guarantees that its solutions can be written as the sum (more generally as an integral) of zonally propagating waves that vary as $A(y)e^{ik(x-Ct)}$ where k is the wave number, C is the phase speed (so its frequency is $\omega = kC$) and $A(y)$ is latitude-dependent amplitude. The amplitude structure function $A(y)$ and the dispersion relation $C(k)$ have yet to be determined. For such zonally propagating wave solutions, system (2.1) becomes:

$$
\begin{bmatrix}
0 & f(y) & \alpha \\
\frac{f(y)}{k^2} & 0 & \frac{\alpha}{k^2}\frac{\partial}{\partial y} \\
1 & -\frac{\partial}{\partial y} & 0
\end{bmatrix}
\begin{bmatrix}
\tilde{u} \\
\tilde{V} \\
\tilde{\eta}
\end{bmatrix}
= C
\begin{bmatrix}
\tilde{u} \\
\tilde{V} \\
\tilde{\eta}
\end{bmatrix}
\tag{2.2}
$$

where $\tilde{V} = \frac{iv}{k}$; $f(y) = \sin\phi_0 + y\cos\phi_0$; and the $(\tilde{u}(y), \tilde{V}(y), \tilde{\eta}(y))$ column vector is the y-dependent amplitudes of u, v and η (the tilde designation of the amplitudes will be omitted from this point onwards since we will refer only to the amplitudes when writing u, V and η).

An examination of the first equation in (2.1) or (2.2), i.e., the x-momentum equation, shows that for traveling wave solutions, it yields the following algebraic relation between u, v and η:

$$
u = \frac{f(y)}{C}V + \frac{\alpha}{C}\eta = \frac{\sin\phi_0 + y\cos\phi_0}{C}V + \frac{\alpha}{C}\eta.
\tag{2.3}
$$

This algebraic (i.e., not differential) relation is a reflection of the fact that the linearized x-momentum equation does not contain y-derivatives.

Substituting the expression for u, (2.3), in the other two equations in (2.2) yields the second-order differential system:

$$
\begin{aligned}
\frac{dV}{dy} &= \frac{\sin\phi_0 + y\cos\phi_0}{C}V + \left(\frac{\alpha}{C} - C\right)\eta, \\
\frac{d\eta}{dy} &= \frac{k^2C^2 - (\sin\phi_0 + y\cos\phi_0)^2}{\alpha C}V - \frac{\sin\phi_0 + y\cos\phi_0}{C}\eta.
\end{aligned}
\tag{2.4}
$$

This system can be transformed to a single, second-order, equation for either V or η by eliminating the other variable. However, in the special case where the constant coefficient of η in the dV/dy equation, $(\alpha - C^2)/C$, vanishes, this elimination cannot be carried out since the equation for V decouples from the η-equation. This special case yields Kelvin waves.

2.1 Kelvin Waves

When $C^2 = \alpha$, the coefficient of η in the dV/dy equation vanishes so V satisfies a first-order equation and the only way for this solution to vanish at the two channel walls, $y = \pm\delta\phi/2$, is for $V(y)$ to vanish everywhere. In this case, η can be solved by setting $V = 0$ in the second equation whose solutions are the non-dimensional counterpart of Kelvin waves in a channel, (1.4), that is,

$$V = 0,$$

$$u = \frac{\alpha}{C}\eta_0 e^{-\frac{(\sin\phi_0 + \cos\phi_0 y)^2}{2C\cos\phi_0}} e^{ik(x-Ct)}, \tag{2.5}$$

$$\eta = \eta_0 e^{-\frac{(\sin\phi_0 + \cos\phi_0 y)^2}{2C\cos\phi_0}} e^{ik(x-Ct)},$$

where $C = \pm\alpha^{1/2}$. As in the dimensional case, the amplitude of the eastward-propagating wave $(C = +\alpha^{1/2})$ decays with the increase in y while that of the westward-propagating wave $(C = -\alpha^{1/2})$ decays with the decrease in y.

2.2 Inertia-Gravity and Planetary Waves

In the general case, when $C^2 \neq \alpha$, one can eliminate either V or η from (2.4) to obtain a single second-order equation. However, since the boundary conditions involve only V it is natural to eliminate η and retain V as the only dependent variable of the sought equation. To eliminate η one differentiates the V-equation with respect to y and employs the original equations in (2.4) to eliminate the terms proportional to $d\eta/dy$ and η from the resulting equation. It turns out that the elimination yields an equation that has no dV/dy term and the resulting second-order equation and its accompanying boundary conditions are:

$$\frac{d^2V}{dy^2} + \left(\frac{\omega^2}{\alpha} - \frac{\cos\phi_0}{C} - k^2 - \frac{(\sin\phi_0 + y\cos\phi_0)^2}{\alpha}\right)V = 0; \quad V\left(y = \pm\frac{\delta\phi}{2}\right) = 0. \tag{2.6}$$

This is a classical eigenvalue problem (equation and boundary conditions) whose solutions are determined by imposing the boundary conditions on the general solutions of the differential equation. The general solution of the equation in (2.6) is a linear combination of parabolic cylinder functions (Abramowitz and Stegun 1972) but there is no explicit way of applying the boundary conditions to this form of the solution. To advance, we first note that $\delta\phi$ appears in the boundary conditions of (2.6) so its role in determining the eigensolutions is more cumbersome to identify compared with the parameters of the differential equation itself.

The change of variables $z = 2y/\delta\phi$ leads to a slightly modified problem in which the boundary conditions are applied at $z = \pm 1$ regardless of the channel width, while the parameter $\delta\phi$ appears in the equation itself instead of in the boundary conditions. In terms of z, (2.6) is written as:

$$\varepsilon^2 \frac{d^2 V}{d^2 z} + \left(E - (\sin \phi_0 + zb)^2 \right) V = 0; \quad V(z = \pm 1) = 0, \tag{2.7}$$

where

$$\varepsilon = \frac{2\sqrt{\alpha}}{\delta\phi}, \quad E = \omega^2 - \frac{\alpha \cos \phi_0}{C} - \alpha k^2 \quad \text{and} \quad b = \frac{\delta\phi}{2} \cos \phi_0. \tag{2.8}$$

The eigenvalue problem (2.7) has the form of a time-independent Schrödinger eigenvalue problem in one dimension in which E is the energy (or eigenvalue to use a more mathematical notion), $f^2 = (\sin \phi_0 + bz)^2$ is the potential and ε is the counterpart of $\hbar(2m)^{-1/2}$ where m is the mass of the particle and $\hbar$ is Planck constant, h, divided by 2π. It has an infinite number of discrete energy values (or levels), E_n, where $n = 0, 1, 2, \ldots$ and each level has an associated eigenfunction, $V_n(z)$, that has exactly n zero-crossings in the $z = (-1, 1)$ interval (i.e., in the inner points excluding the boundaries at the endpoints $z = -1$ and $z = 1$).

In the special case where $\phi_0 = 0$ Eq. (2.7) reduces to the equatorial wave problem so in this Schrödinger eigenvalue problem formulation the equatorial problem is a regular limit of the mid-latitude one, which is the focus of Chap. 3. For general $\phi_0 \neq 0$ Eq. (2.7) can be divided through by $\sin^2 \phi_0$ which results in the mid-latitude eigenvalue problem:

$$\varepsilon^2 \frac{d^2 V}{d^2 z} + \left(E - (1 + zb)^2 \right) V = 0; \quad V(z = \pm 1) = 0, \tag{2.9}$$

where the parameters of this differential equation are

$$\varepsilon = \frac{2\sqrt{\alpha}}{\delta\phi \sin \phi_0}, \quad E = \frac{1}{\sin^2 \phi_0} \left(\omega^2 - \frac{\alpha \cos \phi_0}{C} - \alpha k^2 \right) \quad \text{and} \quad b = \frac{\delta\phi}{2} \frac{\cos \phi_0}{\sin \phi_0}. \tag{2.10}$$

The classical wave theory outlined in the Introduction constitutes the particular $b = 0$ case of (2.9) which is a legitimate mathematical limit but its consistency with the inclusion of $\alpha \cos f_0/C$ in the expression for E in (2.10) is not obvious since the latter term originates from the differentiation of $f(y) = \sin \phi_0 + y \cos \phi_0$ so its presence in E implies that the potential, f^2, is not constant. However, despite its possible inconsistency the $b = 0$ case provides a check on the numerical calculations below and establishes a connection with the classical/harmonic theory. With $b = 0$

(2.9) becomes a constant coefficient equation, so the solution is $V(z) = V_0 \sin\left(\frac{(n+1)\pi}{2}(z+1)\right)$ where V_0 is an arbitrary normalization constant. This special form of the general solution satisfies the boundary conditions at $z = \pm 1$ provided $E_n = 1 + (\varepsilon\pi(n+1)/2)^2$. This particular form of $V(z)$ is precisely the non-dimensional counterpart of the solution for $v(y)$ in (1.6).

Another particular case in which an analytical solution of (2.9) exists is $b = 1$. In this case the potential $(1 + bz)^2 = (1 + z)^2$ vanishes at the equatorward wall, $z = -1$, i.e., the channel extends from the equator to latitude $\delta\phi$. When the $z = (-1, +1)$ domain is doubled to $z = (-3, +1)$ the potential $(1 + z)^2$ becomes a symmetric parabola whose minimum is located at $z = -1$. In this case, (2.9) turns into the well-known eigenvalue equation of Harmonic Oscillator of Quantum Mechanics. The symmetry of solutions of the Harmonic Oscillator is determined by the mode number n, i.e., symmetric solutions have even n and antisymmetric solutions have odd n. Since eigenfunctions of the Harmonic Oscillator are solutions of (2.9) only if they vanish at $z = -1$ the eigenfunctions relevant to (2.9) must be antisymmetric (so they vanish at the mid-point $z = -1$) i.e., those with odd n. The conclusion based on this comparison with the Harmonic Oscillator, where the eigenvalues are $E_m = (2m + 1)\varepsilon$, is that for $b = 1$, the eigenvalues of (2.9) are the eigenvalues of Harmonic Oscillator with odd m, i.e., $m = 2n + 1$ so $E_n = (4n + 3)\varepsilon$.

Numerical solutions of either (2.9) or (2.7) can be obtained straightforwardly using a variety of methods, e.g., shooting methods and finite-difference or collocation methods (the latter methods turn the differential eigenvalue problem into a matrix eigenvalue problem). These numerical methods yield solutions of E_n and V_n for any pair of ε and b values.

An example of such numerical solutions of (2.9) that yield $E_0(\varepsilon, b)$ is shown in Fig. 2.1 (see Paldor et al. 2007). It clearly demonstrates that for large ε the contours are nearly horizontal so E_0 at $b > 0$ is well approximated by its value at $b = 0$. In contrast, at small ε the value of E_0 is very sensitive to the value of b. This finding and the definition of ε in (2.10) suggest that the harmonic theory that applies formally at $b = 0$ yields accurate estimates of the eigensolutions for $b > 0$ only in sufficiently narrow channels where the width, $\delta\phi$, is of the order of the radius of deformation $\alpha^{1/2}\sin\phi_0$.

The definition of E in (2.10) implies that each E_n value is associated with 3 values of C_n given by the roots of the cubic polynomial:

$$k^2 C^2 - \left(E_n \sin^2\phi_0 + \alpha k^2\right) - \frac{\alpha\cos\phi_0}{C} = 0. \tag{2.11}$$

These three roots can be approximated by noting that the absolute value of two of them (one positive and one negative) is large while that of the third root is small. For the small root, we neglect the C^2 term and obtain the approximate expression for the phase speed of the slowly propagating Planetary (Rossby) wave:

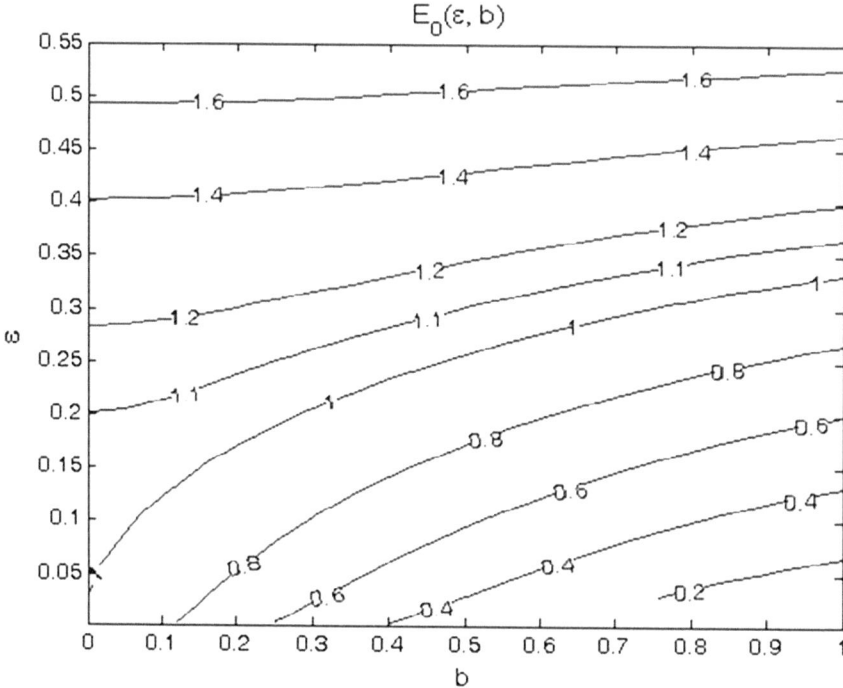

Fig. 2.1 Numerically calculated contours of $E_0(\varepsilon, b)$, the first $(n = 0)$ eigenvalue of (2.9). The analytical solutions $E_n = 1 + (\pi\varepsilon(n+1)/2)^2 \geq 1$ along $b = 0$ and $E_n = (4n + 3)\varepsilon$ along $b = 1$ are fully confirmed by these numerical results. For $0 \ll b < 1$ and for sufficiently small ε the values of E_0 are appreciably smaller than 1. For large ε (e.g., $\varepsilon > 0.5$), E_0 varies only slightly with b. Permission from American Meteorological Society: J. Phys. Oceanogr. doi:10. 1175/JPO2986.1

$$C_n^{\text{Rossby}} \approx \frac{-\cos\phi_0}{k^2 + \frac{\sin^2\phi_0}{\alpha} E_n}. \qquad (2.12)$$

The approximate expression for the two fast Inertia-Gravity waves is obtained by neglecting $\alpha \cos\phi_0 / C$ which yields the approximate expression:

$$\left(C_n^{\text{Poincare}}\right)^2 \approx \alpha + E_n \frac{\sin^2\phi_0}{k^2}. \qquad (2.13)$$

When the explicit approximate dispersion relation of Planetary waves, (2.12), is combined with the effect that b has on the contours of E_0 in Fig. 2.1, it becomes evident that estimates of E_0 based on setting $b = 0$ (where $E_n > 1$ for all ε) are associated with lower phase speeds than those of $b > 0$ (where E_n can be much less than 1). The opposite conclusion holds for Inertia-Gravity waves and (2.13).

It should be noted that while the phase speed of Inertia-Gravity waves in (2.13) differs from that of Planetary waves in (2.12) the eigenfunction $V(y)$ in (2.9) of the two waves is identical because the two waves belong to the same eigensolution of the eigenvalue equation. Since the expressions that relate $u(y)$ and $\eta(y)$ to $V(y)$ and dV/dy involve the phase speed, C (i.e., (2.3) for u and the first equation in (2.4) for η), these functions are not identical in the two waves.

Before turning to an analysis of the eigenfunctions V_n associated with the eigenvalues E_n we turn our attention to Inertial waves that do not exist in the harmonic theory.

2.3   Inertial Waves

The difference between the values of E_0 for $b = 0$, where $E_0 > 1$ and $b > 0$, where E_0 can be smaller than 1 brings about a fundamental difference between these two cases when it comes to Inertial waves. As explained in the Introduction, these waves should be thought of as a limiting case of Inertia-Gravity waves when no pressure gradient forces act on the fluid, i.e., when the total wave number, $k^2 + n^2$, vanishes in (1.5). In the harmonic, $b = 0$, theory these waves cannot satisfy the two boundary conditions at the channel walls (see the discussion at the end of Sect. 3.9 in Pedlosky 1987) and they are therefore regarded as "spurious solutions" of the differential equations.

As implied by their names the frequency of Inertial waves is the Coriolis frequency, f_0, so in the present non-dimensional formulation becomes $\omega^2 = \sin^2 \phi_0$ and in this case (2.10) implies that $\sin^2 \phi_0 (1 - E) = \alpha k (k + \cos \phi_0 / \omega)$. Therefore, for $k = 0$ these waves can exist only when $E = 1$ while in the case where $k + \cos \phi_0 / \omega > 0$ (i.e., either $\omega = + \sin \phi_0$ or $\omega = - \sin \phi_0$ and $k > \cot \phi_0$) E should be smaller than 1 so Inertial waves cannot exist for $E > 1$. In the $b = 0$ case, $E_n = 1 + (\pi \varepsilon (n + 1))^2 / 4 > 1$ and therefore no Inertial waves can exist in this case and the only solution of eigenvalue problem (2.9) is the trivial $V = 0$ which is the foundation of the conclusion reached in Pedlosky (1987). In contrast, Fig. 2.1 clearly shows that contours with $E_0 \leq 1$ do exist provided $b > 0$. Thus, in terms of the eigenvalue problem (2.9) Inertial waves with either $k = 0$ or $k > 0$ are "spurious solutions" only for $b = 0$, i.e., in the harmonic theory but these waves are legitimate solutions of the Trapped wave theory where $E_0 \leq 1$ eigenvalues exist for $b > 0$.

2.4 Analytic Solutions of the Eigenvalue Problem and Eigenfunctions

In order to obtain analytic insight into the structure of the eigenfunctions and in order to derive explicit expression for E_n when $b > 0$ we note that in the β-plane approximation second-order terms in $f(y)$ (i.e., second-order terms in the expansion of $\sin \phi$ near ϕ_0) are neglected. Therefore, terms of order y^2 should be neglected in the potential of (2.6), i.e., the b^2z^2 term should be neglected in the potentials of (2.7) and (2.9) for these equations to be consistent with the neglect of second-order terms in the expansion of $f(y)$ on the β-plane. The neglect of b^2z^2 is consistent with the restriction on values of $b\ (= \beta L/f_0 = L \cot \phi_0/2a)$ which has to be smaller than 1 but larger than 0. Since eigensolutions of two Schrödinger eigenvalue problems with close potentials are always close to one another analytic insight can be gained by approximating the potential in (2.6) by a linear function in y (and that of (2.9) by a linear function of z).

When b^2z^2 is neglected in (2.9) this equation becomes:

$$\varepsilon^2 \frac{d^2 V}{d^2 z} + (E - (1 + 2zb))V = 0; \quad V(z = \pm 1) = 0, \tag{2.14}$$

and the parameters in this equation maintain their original definitions, (2.10). The elimination of b^2z^2 yields an equation that can be transformed (by dividing Eq. (2.15) through by ε^2 and defining a new independent variable) to have only two parameters, $(E - 1)/\varepsilon^2$ and $2b/\varepsilon^2$, instead of three (ε, E and b) in (2.9). This reduction in the number of parameters that determine the solution of the equation greatly simplifies the analysis of these solutions.

The linear potential in (2.14) suggests that the differential equation can be transformed to an Airy equation (see Sect. 10.4 in Abramowitz and Stegun 1972). As was shown in Paldor and Sigalov (2008) transforming the independent variable z to new independent variable Z defined by

$$Z = (4b^2\varepsilon^2)^{-1/3}(2bz - E + 1) \tag{2.15}$$

transforms the differential Eq. (2.14) to Airy equation

$$\frac{d^2 V}{dZ^2} - ZV = 0. \tag{2.16}$$

As a solution of a second-order differential equation, $V(Z)$ in (2.16) can be expressed as a linear combination of two independent solutions: $Ai(Z)$ and $Bi(Z)$. One can get a rough idea on the qualitative behavior of Airy functions by replacing the coefficient Z in front of V in (2.16) by a constant in which case the behavior of the solutions of the modified, constant coefficient, equation is well known. When the "constant" Z is positive the solutions decay/grow while when it is negative the

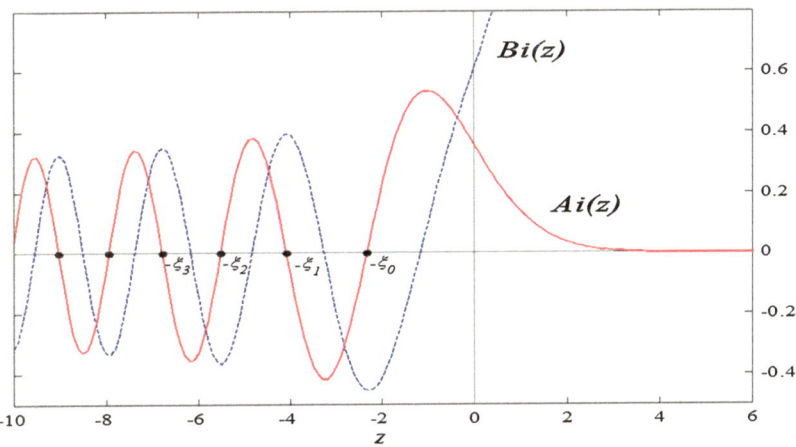

Fig. 2.2 The two independent solutions of Airy equation, (2.16), Ai(Z) and Bi(Z). While Bi (Z) grows faster than exponential in the $Z > 0$ half-plane Ai(Z) decays to zero there faster than exponential. Both functions oscillate a-periodically in the $Z < 0$ half-plane

solutions oscillate periodically. Similarly, at $Z > 0$ Ai(Z) decays while Bi(Z) grows both at a rate faster than exponential and at $Z < 0$, both solutions oscillate but are not periodic. The zeros of Ai and Bi can be easily found either in tables such as that given in Abramowitz and Stegun (1972) or by resorting to readily available mathematical packages such as MATLAB or Mathematica. A sketch of these two solutions is shown in Fig. 2.2.

The final step that completes the transformation of the differential equation is the application of the boundary conditions. To satisfy the boundary condition V $(z = -1) = 0$, the point $Z(z = -1)$ has to be set to the nth zero of Ai, $-\xi_n$, which ensures that $V(Z(z = -1; E = E_n)) = \text{Ai}(-\xi_n) = 0$. The requirement $Z(z = -1; E = E_n) = -\xi_n$ in the transformation: $Z = (4b^2\varepsilon^2)^{-1/3}(2bz - E + 1)$ yields $-\xi_n = (4b^2\varepsilon^2)^{-1/3}(-2b - E_n + 1)$ which can be inverted to the following explicit expression for the nth eigenvalue:

$$E_n = 1 - 2b + (2b\varepsilon)^{2/3}\xi_n = 1 - \delta\phi\cot\phi_0 + \left(2\frac{\cos\phi_0\sqrt{\alpha}}{\sin^2\phi_0}\right)^{2/3}\xi_n \qquad (2.17)$$

where ξ_n is the absolute value of the nth zero of Ai. The first five ξ_n's are $\xi_0 = 2.3381$, $\xi_1 = 4.08795$, $\xi_2 = 5.5205$, $\xi_3 = 6.7867$ and $\xi_4 = 7.9441$ and while the value of ξ_n increases with n the difference between two successive zeros of Ai $\xi_n - \xi_{n+1}$ decreases with n.

To satisfy the boundary condition at the poleward wall, $V(z = +1) = 0$, the value of Z at $z = +1$ has to be sufficiently large (and positive) such that Ai(Z(z = 1)) is small enough to be regarded as 0. This boundary condition limits the applicability of the present theory to wide channels only. Since Ai(2) ≈ 0.035 and since

$Ai(2)/Bi(2) \leq 0.01$, the choice $Z(z=1) \geq 2$ guarantees both that V is negligibly tiny at the poleward wall and that the contribution of Bi, the exponentially growing solution of Airy equation, is uniformly negligible throughout the entire channel (the negligible contribution of Bi is an essential element in the determination of E_n based on the location of the zeros of Ai only). Substituting $Z \geq 2$ in the general $Z(z)$ transformation given in Eq. (2.15) $Z = (4b^2\varepsilon^2)^{-1/3}(2bz - E + 1)$, with $z = 1$ and $E = E_n$ (where E_n is given by (2.17)) yields:

$$\xi_n + 2 \leq 4b(2b\varepsilon)^{-2/3} = 4^{2/3}b^{1/3}\varepsilon^{-2/3}. \tag{2.18}$$

Substituting ε and b from (2.10) into (2.18) yields, $\xi_n + 2 \leq \delta\phi \left(\frac{2\sin\phi_0 \cos\phi_0}{\alpha} \right)^{1/3}$ which can be inverted to yield the following lower bound on the value of the channel width, $\delta\phi$, above which the Trapped wave theory applies:

$$\delta\phi \geq (\xi_n + 2)\left(\frac{\alpha}{\sin 2\phi_0} \right)^{1/3}. \tag{2.19}$$

According to the values of the first five ξ_n's given above, the coefficient $(\xi_n + 2)$ on the RHS of (2.19) varies between 4.3 for $n = 0$ and 9.9 for $n = 4$. At $n = 20$, $\xi_{20} + 2$ is 22 and at $n \geq 20$ $\xi_{n+1} - \xi_n$, the difference between successive zeros of Ai, is less than 0.5 (and it decreases further with n). Thus, with an error of no more than a factor of 2 and for not-too-large mode number, n, one can set $(\xi_n + 2)$ to 10. For ϕ_0 in the mid-latitudes $\sin^{1/3} 2\phi_0$ is very close to 1 as $\sin^{1/3} 2\phi_0$ varies between 0.95 and 1 when ϕ_0 varies between 30° and 75°. Consequently, the present theory is valid for channel widths satisfying $\delta\phi \geq 10\alpha^{1/3}$ so for a baroclinic ocean where $\alpha \approx 5 \times 10^{-6}$ the present theory is valid provided $\delta\phi > 0.17$ rad, i.e., less than 10°, or 1000 km, of latitude. This is close to the latitudinal range over which the β-plane approximation is valid—at larger ranges, the linearity of $f(y)$ cannot be justified while at lower ranges, the f-plane approximation (where $f(y)$ is assumed constant) is valid. In contrast, in a barotropic ocean (and atmosphere) where $\alpha \approx 5 \times 10^{-2}$ the channel width has to satisfy $\delta\phi > 1$ rad even for $n = 0$, which is much larger than the range of validity of the β-plane approximation. The conclusion from this discussion is that the present Trapped wave theory is applicable to a baroclinic ocean on the β-plane.

Similar arguments can be made for delineating the regime of channel width where Harmonic waves can be expected to prevail. An upper bound on the channel width below which Harmonic wave theory is valid can be estimated by noting that if the poleward wall, $z = 1$, is located at $Z = 0$ then, as is evident from Fig. 2.2, the two Airy solutions oscillate throughout the entire channel and in such an oscillatory regime the purely oscillatory Harmonic waves provide an accurate solution. For the $n = 0$ mode and at 45°, (2.19) yields $\delta\phi \geq 4.3\alpha^{1/3}$ as the lower bound on the channel width above which Trapped waves can be expected to prevail while for $\delta\phi \leq 2.3\alpha^{1/3}$ Harmonic waves should prevail. Thus, in a barotropic ocean where

$\alpha = 0.05$, the first Harmonic mode prevails at all physically acceptable widths (i.e., smaller than 0.85 rad $\approx 50°$ so the channel extends from above 20° to below 70°), while in a baroclinic ocean where $\alpha = 5 \times 10^{-6}$ the first Trapped mode prevails at all channels with widths exceeding 0.0735 rad, i.e., wider than about 4°, which is hardly the range that justifies the application of the β-plane approximation (the f-plane suffices).

For values of α and $\delta\phi$ satisfying (2.19) the dispersion relations of Planetary and Inertia-Gravity waves are obtained by substituting (2.17) in (2.12) and (2.13), respectively:

$$\omega_n^{Rossby} \approx \frac{-\cos\phi_0 k}{k^2 + \frac{\sin^2\phi_0}{\alpha}E_n} = \frac{-\cos\phi_0 k}{k^2 + \frac{\sin^2\phi_0}{\alpha} - \frac{\delta\phi}{2}\frac{\sin 2\phi_0}{\alpha} + \alpha^{-2/3}\xi_n(\sin 2\phi_0)^{2/3}} \quad (2.20)$$

and

$$\left(\omega_n^{Poincare}\right)^2 \approx \alpha k^2 + \sin^2\phi_0 E_n = \sin^2\phi_0 + \alpha k^2 - \frac{\delta\phi}{2}\sin 2\phi_0 + \alpha^{1/3}\xi_n(\sin 2\phi_0)^{2/3}. \quad (2.21)$$

These dispersion relations should be compared with their classical dimensional counterparts, (1.5) for Planetary waves and (1.7) for Inertia-Gravity waves. The comparison clarifies that in both waves the arbitrary (i.e., independent of the radius of deformation) meridional wave number of the harmonic theory, $n = (2N + 1) \cdot \pi/L$ (where L is the channel width) is replaced in the Trapped wave theory by two terms: The first terms is proportional to the channel width, $\delta\phi$, and the second term (proportional to $\alpha^{-2/3}$ in the case of planetary waves) has the coefficient ξ_n which determines the meridional mode number, n, as the number of zeros that $V(y)$ crosses inside the channel, i.e., the mode number and channel width appear in different terms of the dispersion relation instead of $nL = $ constant in the Harmonic theory.

Numerical solutions of the eigenvalue problem associated with zonally propagating waves in a mid-latitude channel (including Kelvin, Planetary, and Inertia-Gravity waves) can be obtained by solving system (2.2) numerically. A powerful method of solution of such differential-matrix eigenvalue system is the Chebyshev spectral collocation method on a grid (see e.g., Trefethen (2000), Poulin and Flierl (2003) and De-Leon and Paldor (2009) for more details) in which the (u, V, η) system, (2.2), is solved as a matrix eigenvalue problem of linear algebra by transforming the differential $\partial/\partial y$ operators to an algebraic form (as in the case when derivatives are written as differences). The eigenvalues of this matrix eigenvalue problem are the phase speeds, C, while the corresponding eigenvectors consist of the three amplitudes $u(y)$, $V(y)$, and $\eta(y)$.

A numerical solution of the eigenvectors in a channel with $b = 0.15$ and $\varepsilon = 0.055$ (so the channel width, $\delta\phi$, is 36 deformation radii, $\alpha^{1/2}/\sin\phi_0$) where the present theory applies is presented in Fig. 2.3, reproduced from Paldor et al. (2007). Consistent with the analytical conclusions reached above, the $V(y)$ eigenfunction is

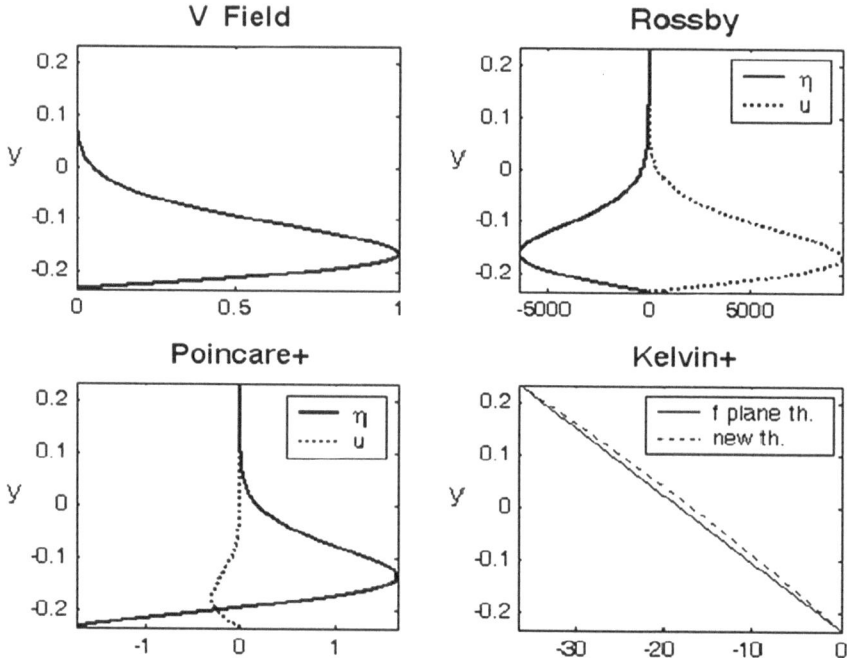

Fig. 2.3 The numerically calculated y-dependent amplitude eigenfunctions of the three wave types in a wide channel ($\varepsilon = 0.055$, i.e., the width, $\delta\phi$, equals 36 deformation radii, $\alpha^{1/2} / \sin \phi_0$). Ordinates are the cross-channel coordinate: $y = z \cdot \delta\phi$ and abscissas are arbitrary amplitudes (normalized in Rossby and Poincaré waves such that max$\{V\} = 1$). Permission from American Meteorological Society: J. Phys. Oceanogr. doi:10.1175/JPO2986.1

identical in Planetary and Inertia-Gravity waves (recall that there is a single V (y) eigenfunction to (2.14) or (2.16)), while the $u(y)$ and $\eta(y)$ of the former waves differ from those of the latter because of the different phase speeds of the two wave types i.e., the values of C given in (2.12) and (2.13) affect the relation between η and V and dV/dy in (2.4) and that between u and V and η in (2.3). The trapping of V (y) near the equatorward wall anticipated from the behavior of the regular Airy function is clearly evident in these numerical results which cannot be recovered by the harmonic theory outlined in the Introduction since $E_0 = 0.862$, whereas in the harmonic theory, $E_0 = 1.00187$ in this case.

  In contrast, when the channel is sufficiently narrow, the Trapped wave theory does not apply and one can expect the theory outlined in the Introduction to apply. An example for this case is shown in Fig. 2.4 which repeats the calculation shown in Fig. 2.3 but for a channel width that equals the deformation radius. As in Trapped waves, the V eigenfunctions are identical for Planetary and Inertia-Gravity waves, but the u and η functions are not. The main feature of the eigenfunctions in this case is that they spreads all over the channel and are not trapped near the equatorward wall. The theory outlined in the Introduction prevails in this case since $E_0 = 10.871$

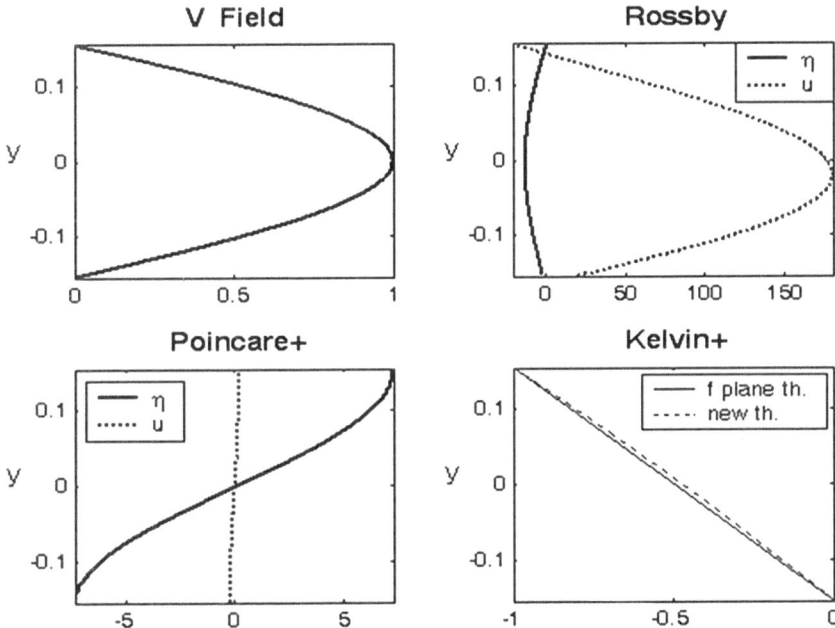

Fig. 2.4 The y-dependent amplitude eigenfunctions in a narrow, $\varepsilon = 2$, channel (i.e., the channel width, $\delta\phi$, equals the deformation radius, $\alpha^{1/2}/\sin\phi_0$). Ordinates, abscissas and normalizations are identical to those in Fig. 2.3. Permission from American Meteorological Society: J. Phys. Oceanogr. doi:10.1175/JPO2986.1

where the value of b does not affect the solution according to the results shown in Fig. 2.1, so the $b = 0$ results yield accurate estimates for all values of b. These results are relevant to a channel width of $\delta\phi = 0.3$ only for barotropic ocean with $\alpha = 0.05$, while in a baroclinic ocean where $\alpha = 5 \times 10^{-6}$ and at $45°$ latitude, the channel width where this mode exists is only 0.003 rad or about $0.17°$ (a dimensional width of less than 20 km!). The numerical solutions underscore the inaccuracy of the theory outlined in the Introduction for Rossby waves since the $u(y)$ eigenfunction of Planetary (Rossby) waves does not vanish at the channel walls in contradiction to the expressions given in (1.8) (note that $V(y)$ in Eq. (2.2) and in Figs. 2.3 and 2.4 is shifted by $\pi/2$ relative to $v(y)$ in (1.8) so $V(y)$ has the same y-dependence as $u(y)$).

To summarize this chapter that focuses on waves in a mid-latitude channel on the β-plane it is instructive to compare the phase speed of Trapped Planetary (Rossby) waves, (2.20), with that of Harmonic Planetary waves (1.7). Dividing (1.7) by $2\Omega ak$ yields an expression for the non-dimensional Harmonic phase speed C_{Harm} which can be compared with the phase speed of Trapped waves, C_{Trap}, obtained by dividing (2.20) by k. To enable the comparison, we fix the central latitude, ϕ_0, to $45°$ (so $\sin 2\phi_0 = 1$ and $\sin^2 \phi_0 = 1/2$) and assign the radius of deformation, $R_d = (gH)^{1/2}/f_0$, the typical baroclinic value of 25 km so $(gH)^{1/2}/2\Omega = 17.7$ km and $\alpha = (17.7/6400)^2 \approx 7.6 \times 10^{-6}$. The zonal wave number, k, of a wave with

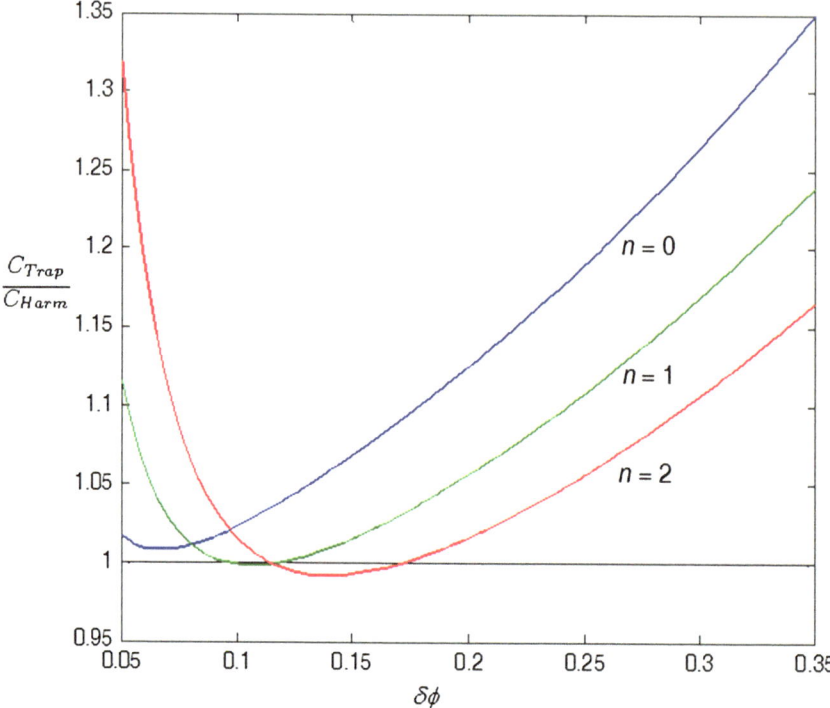

Fig. 2.5 The ratio between the phase speeds of the first three Trapped modes to those of the corresponding Harmonic modes

640 km wavelength, has a non-dimensional value of $k = 2\pi\,6400/640 = 20\pi$ so $k^2 \approx 4 \times 10^3 \ll \alpha^{-1} \approx 1.5 \times 10^5$. For these values of ϕ_0, k^2 and α, the resulting expression for the ratio between Trapped and Harmonic phase speeds of Planetary waves is

$$
\begin{aligned}
\frac{C_{\text{Trap}}}{C_{\text{Harm}}} &= \frac{k^2 + \frac{\sin^2\phi_0}{\alpha} + \frac{(n+1)^2\pi^2}{(\delta\phi)^2}}{k^2 + \frac{\sin^2\phi_0}{\alpha} + \left(\xi_n \cdot \left(\frac{\sin 2\phi_0}{\alpha}\right)^{2/3} - \frac{\delta\phi}{2}\frac{\sin 2\phi_0}{\alpha}\right)}, \\[2mm]
&\approx \frac{1 + 2\alpha\frac{(n+1)^2\pi^2}{(\delta\phi)^2}}{1 + 2\xi_n\alpha^{1/3} - \delta\phi}
\end{aligned}
\tag{2.22}
$$

where the index $n = 0, 1, 2\ldots$ is identical in the numerator and denominator (i.e., n was increased by 1 relative to its value in (1.7) so as to match the values of n in (2.17) and (2.20)).

The phase speed ratio, (2.22), is plotted in Fig. 2.5 as a function of the channel width $\delta\phi = L/a$ for $n = 0$, 1, and 2 and it shows that the phase speed of Trapped waves is larger than that of Harmonic waves in sufficiently wide channels. For $\phi_0 = \pi/4$ and $\alpha = 7.6 \times 10^{-6}$, the right-hand side (RHS) of (2.19) is very close to $0.02 \cdot (\xi_n + 2)$ so for $n = 0, 1$, and 2, the Trapped wave theory applies to channels of widths exceeding 0.087, 0.12, and 0.15 rad, respectively, consistent with the location of the minima of the three curves shown in Fig. 2.5. The monotonic increase of the three curves in Fig. 2.5 at $\delta\phi$ larger than the minimal widths and the fact that in their ascending branches, the ratios are larger than 1 suggest that when wide channels are used to approximate an infinite ocean, the phase propagation of Trapped Planetary waves is appreciably faster than that of Harmonic waves, exceeding a factor 2 in very wide channels.

Having completed the derivation of the Trapped wave theory in a channel on the mid-latitude β-plane it is quite natural to examine the same problem in a channel on the equatorial β-plane as a limiting case of the mid-latitude theory when ϕ_0 is set equal to 0. This is the subject of the next chapter.

References

Abramowitz M, Stegun IA (1972) Handbook of mathematical functions. Dover Publications Inc, USA, p 1043

De-Leon Y, Paldor N (2009) Linear waves in mid-latitudes on the rotating spherical earth. J Phys Oceanogr 39:3204–3215

Paldor N, Rubin S, Mariano AJ (2007) A consistent theory for linear waves of the shallow water equations on a rotating plane in mid-latitudes. J Phys Oceanogr 37(1):115–128

Paldor N, Sigalov A (2008) Trapped waves on the mid-latitude β-plane. Tellus 60A: 742–748. doi:10.1111/j.1600-0870.2008.00332.x

Pedlosky J (1987) Geophysical fluid dynamics. Springer, Berlin, 710 p

Poulin FJ, Flierl GR (2003) The nonlinear evolution of barotropically unstable jets. J Phys Oceanogr 33(10):2173–2192

Trefethen LN (2000) Spectral methods in MATLAB. SIAM, 165 p

Chapter 3
Waves in a Channel on the Equatorial β-Plane

Having found both Harmonic and Trapped waves in a mid-latitude channel it is natural to ask why is it that in an equatorial channel only Trapped waves (the Hermite function waves that are trapped to the equator) were found in Matsuno (1966) while Harmonic waves have not been found there. A related question is what type of solutions exists in an equatorial channel in view of the fact that Matsuno's theory was developed on the unbounded equatorial β-plane which is inconsistent with both the neglect of higher-order terms in the expansion of $\sin\phi$ and the planar approximation to the spherical earth. In the present formulation an equatorial channel is simply addressed by setting $\phi_0 = 0$ in (2.7)—the eigenvalue equation in a channel on the mid-latitude β-plane—which transforms this equation to:

$$\varepsilon^2 \frac{d^2 V}{d^2 z} + \left(E - (zb)^2 \right) V = 0; \quad V(z = \pm 1) = 0, \tag{3.1}$$

where the parameters ε, E, and b are defined by setting $\cos\phi_0 = 1$ in (2.8); i.e.,

$$\varepsilon = \frac{2\sqrt{\alpha}}{\delta\phi}, \quad E = \omega^2 - \frac{\alpha}{C} - \alpha k^2 \quad \text{and} \quad b = \frac{\delta\phi}{2}. \tag{3.2}$$

The eigenvalue problem (3.1) can be simplified so as to combine the three parameters E, ε, and b to two parameters by dividing the differential equation through by ε^2 which does not affect the boundary conditions. The eigenvalue problem then becomes:

$$\frac{d^2 V}{d^2 z} + \left(\frac{E}{\varepsilon^2} - \left(\frac{b}{\varepsilon} \right)^2 z^2 \right) V = 0; \quad V(z = \pm 1) = 0, \tag{3.3}$$

in which E/ε^2 is determined by the single parameter of the potential b/ε and the two are defined by:

$$\frac{E}{\varepsilon^2} = \frac{(\delta\phi)^2}{4\alpha} \left(\omega^2 - \frac{\alpha}{C} - \alpha k^2 \right) \quad \text{and} \quad \frac{b}{\varepsilon} = \left(\frac{\delta\phi}{2} \right)^2 \frac{1}{\sqrt{\alpha}}. \tag{3.4}$$

© The Author(s) 2015
N. Paldor, *Shallow Water Waves on the Rotating Earth*,
SpringerBriefs in Earth System Sciences, DOI 10.1007/978-3-319-20261-7_3

The differential Eq. (3.3) is similar to that studied in Matsuno (1966) but in Matsuno's work the boundary conditions were applied at infinity. The modified "wall" problem was studied by Erlick et al. (2007) who showed that with this change in boundary conditions, $(b/\varepsilon)^2$ can be eliminated from the potential by transforming the independent variable from z to $(b/\varepsilon)^{1/2}z$ and dividing the equation through by (b/ε) (so the eigenvalue changes from E/ε^2 to $E/(b\varepsilon) = E/\alpha^{1/2}$ which makes the problem independent of $\delta\phi$). However, this transformation filters out the narrow channel solution derived next based on the boundness of the domain in (3.3), where $-1 \leq z \leq 1$ (i.e. $z^2 \leq 1$).

According to the definitions of the parameters given in (3.4), the narrow channel solution is relevant in the limit $b/\varepsilon \propto (\delta\phi)^2 \to 0$. When terms of order $(b/\varepsilon)^2$ are neglected in (3.3), the potential $(bz/\varepsilon)^2$ vanishes and the equation becomes a constant coefficient equation whose solution is $V(z) = A\sin(E^{1/2}z/\varepsilon + \theta)$ where A is an arbitrary normalization amplitude and $E^{1/2}/\varepsilon$ and θ are determined by the boundary conditions $V(z = \pm1) = 0$; i.e.,

$$V_n(z) = A\sin\left(\frac{(n+1)\pi}{2}(z+1)\right), \quad n = 0, 1, 2, \ldots \tag{3.5}$$

and the eigenvalues are given by:

$$\frac{E_n^{1/2}}{\varepsilon} = \frac{(n+1)\pi}{2} \Leftrightarrow E_n = \left(\frac{(n+1)\pi}{\delta\phi}\right)^2 \alpha, \quad n = 0, 1, 2, \ldots \tag{3.6}$$

The second solution is encountered when $b/\varepsilon \propto (\delta\phi)^2$ is O(1) in which case the differential equation in (3.3) is the known equation of Harmonic Oscillator of Quantum Mechanics (see, e.g., Schiff 1968). However, the boundary conditions in the latter problem are applied at $y = \pm\infty$ whereas in (3.3) they are applied at $z = \pm1$. The slightly different boundary conditions in (3.3) will be handled shortly but for now we note that the solutions of the Harmonic Oscillator differential equation in (3.3) are Hermite functions:

$$V_n(z) = A \cdot H_n\left(\frac{\delta\phi}{2}\left(\frac{1}{\alpha}\right)^{1/4}z\right) \cdot e^{-\left(\frac{\delta\phi}{2}\right)^2\frac{1}{\sqrt{\alpha}}\frac{z^2}{2}}, \quad n = 0, 1, 2, \ldots \tag{3.7}$$

where H_n is the Hermite polynomial of order n and A is an arbitrary normalization constant. The eigenvalues associated with the eigenfunctions (3.7) are:

$$\frac{E_n}{\varepsilon^2} = (2n+1)\frac{b}{\varepsilon} \Leftrightarrow E_n = (2n+1)b\varepsilon = (2n+1)\alpha^{1/2}. \tag{3.8}$$

In order for the Hermite functions in (3.7) and the eigenvalues in (3.8) to be an eigensolution of the eigenvalue problem (3.3) the eigenfunctions have to vanish at walls, $z = \pm 1$. Regardless of the value of n, the solution for V in (3.7) vanishes at the walls $z = \pm 1$ provided the exponent on the RHS of (3.7), $b/(2\varepsilon)$, is larger than, say, 3 since in this case $V(z = \pm 1)/V(0) \le e^{-3} \approx 0.04 \ll 1$; i.e., the value of V at the walls is two orders of magnitude smaller than its maximal value at the equator, $z = 0$. The definition of b/ε in (3.4) implies that the condition $b/(2\varepsilon) > 3$ is equivalent to $(\delta\phi)^2 > 24\alpha^{1/2}$, which is satisfied when $\delta\phi \ge 5\alpha^{1/4}$.

The exact expressions for the eigenvalues (3.6) and (3.8) in narrow and wide channels, respectively, can be combined to a simple but fairly accurate parameterized expression for all channel widths that degenerates accurately to the exact expressions for $b/\varepsilon \ll 1$ and $b/\varepsilon \gg 1$.

$$\frac{E_n}{\varepsilon^2} = \left[\left(\frac{(n+1)^2 \pi^2}{4} \right)^3 + \left((2n+1)\frac{b}{\varepsilon} \right)^3 \right]^{1/3}, \tag{3.9}$$

which yields the explicit solution for the eigenvalues E_n when b and ε are substituted from (3.2):

$$E_n = \left((2n+1)^3 \alpha^{3/2} + \left(\frac{(n+1)^2 \pi^2 \alpha}{(\delta\phi)^2} \right)^3 \right)^{1/3}. \tag{3.10}$$

Figure 3.1 shows the first three eigenvalues, E_0/ε^2, E_1/ε^2, and E_2/ε^2, obtained from numerical solutions of (3.3) as functions of b/ε. The $n = 2$ curve is also compared with the parameterized expression (3.10). The numerical solutions also agree very well with the asymptotic solutions (3.6) and (3.8) at small and large b/ε, respectively and an inspection of the curves in this figure shows that these analytic expressions provide very accurate approximations to the eigenvalues in their respective range of applicability.

Focusing on the $n = 0$ mode, where the Hermite function is a Gaussian, the results shown in Fig. 3.2 demonstrate that, as anticipated earlier based on the rate of decay of the eigenfunction, when $b/\varepsilon > 6$, the error by the wide channel solution is of order 0.01 (and decreasing with further increase of b/ε) while the accuracy of the narrow channel solution is a few percent when $b/\varepsilon < 1$. Thus, the "infinitely" wide channel approximation employed by Matsuno (1966) applies when $b/\varepsilon = (\delta\phi)^2/(4\alpha^{1/2}) > 6$; i.e., $\delta\phi \ge 5\alpha^{1/4}$. Since the equatorial β-plane approximation applies for $\delta\phi < 0.5$ Rad (i.e., when the channel walls are located at about ± 0.25 Rad $\approx \pm 15°$), the validity of the wide channel solution is limited to $\alpha \le (0.5/5)^4 = 10^{-4}$ which is 500 times smaller than the typical value in a barotropic ocean, $\alpha = 5 \times 10^{-2}$. The definition of $\alpha = (C_g/2\Omega a)^2$ (where C_g is the speed of gravity waves) implies therefore that Matsuno's unbounded equatorial β-plane theory does not apply to fluids in which $C_g \ge 10$ m/s.

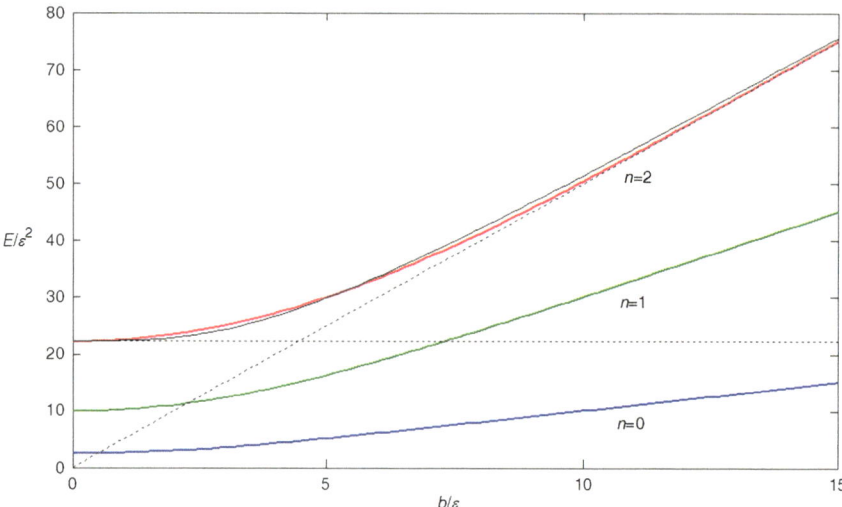

Fig. 3.1 Numerical solutions of Eq. (3.3) and the analytic asymptotic relationships between E/ε^2 and b/ε—(3.6) and (3.8), shown by the *straight dashed lines* for $n = 2$. The parameterized global approximation (3.10) for $n = 2$ is also shown by the *thin solid line* near the $n = 2$ curve

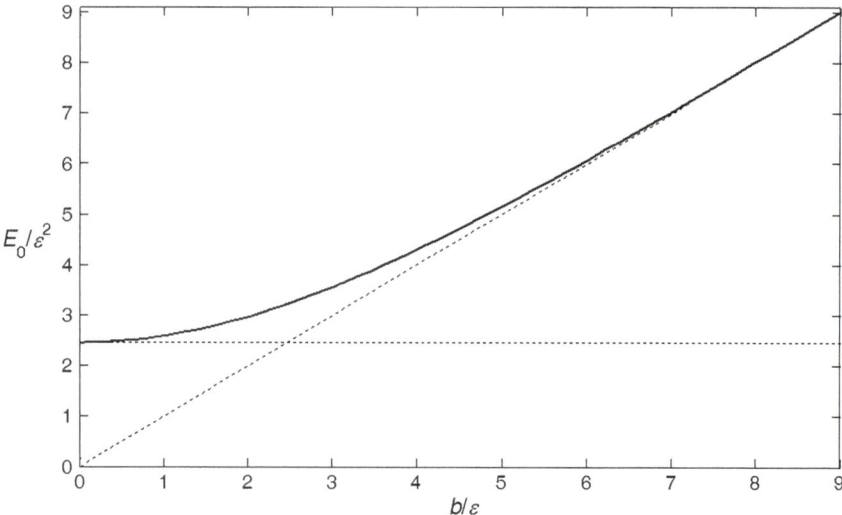

Fig. 3.2 The dependence of E_0/ε^2 on b/ε and the two asymptotic approximations (3.6) and (3.8) for *narrow* and *wide channels*. The error of the wide channel solution is small to within 1 % only when $b/\varepsilon \geq 6$

The explicit expressions derived above for the energy, E_n, as a function of α and $\delta\phi$ can now be applied to obtain the dispersion relations, $C(k)$ (or $kC(k)$), of Planetary and Inertia-Gravity waves from the cubic $E(C)$ relation in (3.2).

Substituting the wide channel expression for E_0 given in (3.8) into the cubic $\omega_n(E_n)$ relation (3.2) (recall: $\omega = kC$) yields:

$$\omega_n^2 - \frac{\alpha k}{\omega_n} - \alpha k^2 = \alpha^{1/2}(2n+1). \tag{3.11}$$

Up to a factor of $\alpha^{1/2}$ and accounting for the different sign of k or ω (Matsuno assumed that the waves vary as $e^{i(kx+\omega t)}$ while here the waves are assumed to vary as $e^{i(kx-\omega t)}$) this equation is identical to Matsuno's Equation (8). Consequently, all of Matsuno's results that were derived on the unbounded equatorial β-plane apply to the wide channel asymptote, $\delta\phi \geq 5\alpha^{1/4}$, of the channel theory. In the limits of large ω_n and small ω_n in (3.11), the approximate expressions of the dispersion relations are as follows:

$$\omega_n^2 = \alpha k^2 + \alpha^{1/2}(2n+1) \tag{3.12}$$

for Inertia-Gravity waves and

$$\omega_n = -\frac{\alpha k}{\alpha k^2 + \alpha^{1/2}(2n+1)} \tag{3.13}$$

for Planetary waves. The $n = 0$ case is unique in that the high-frequency negative root of (3.12) can be close to the low-frequency root of (3.13) for certain combinations of α and k so the approximate expressions (3.12) and (3.13) do not approximate the values of these frequencies accurately. However, for $n = 0$ the exact cubic equation in (3.11) can be decomposed to:

$$0 = \omega_0^3 - \omega_0\left(\alpha k^2 + \alpha^{1/2}\right) - \alpha k = \left(\omega_0 + \alpha^{1/2}k\right)\left(\omega_0^2 - \alpha^{1/2}k\omega_0 - \alpha^{1/2}\right). \tag{3.14}$$

The first root of this special $n = 0$ polynomial $\omega_0 = -\alpha^{1/2}k$ (i.e. $C = -\alpha^{1/2}$) is the "anti-Kelvin" mode, i.e., the westward propagating Kelvin wave which is one of the singular cases where the V equation of the (V, η) system, (2.4), decouples from the η equation and the only solution that satisfies the boundary conditions at the channel walls is $V = 0$ everywhere. Setting $\phi_0 = 0$ in (2.5) clarifies that in this $V = 0$ case with $C = -\alpha^{1/2} < 0$, the solution for $\eta(y)$, $e^{-\frac{y^2}{2C}}$, increases fast with distance from the equator and according to (2.3), when $V = 0$, so does the meridional variation of u. Due to its rapid growth with y this root is a singular solution on the unbounded β-plane as well as in the wide channel asymptote of the channel theory but is physically acceptable in the narrow-channel asymptote of the channel theory.

The other two roots of (3.14) are as follows:

$$(\omega_0)_{1,2} = \frac{\alpha^{1/2}}{2}\left(k \pm \sqrt{k^2 + 4\alpha^{-1/2}}\right). \tag{3.15}$$

The positive root, $(\omega_0)_1$, is the eastward propagating Inertia-Gravity wave whose frequency tends to $+\alpha^{1/4}$ when $k \to 0$ and to $+\alpha^{1/2}k$ for $k \to \infty$. The negative root, $(\omega_0)_2$, of this equations is the mixed mode that has a negative, nonzero, value of $-\alpha^{1/4}$ when $k \to 0$ as in the westward propagating Inertia-Gravity modes (where the dispersion curve intersects the $k = 0$ ordinate at nonzero value) and tends to 0 when $k \to \infty$ as in Planetary waves. By differentiating $(\omega_0)_2$ with respect to k, it can be easily verified that the increase of $(\omega_0)_2$ from $-\alpha^{1/4}$ at $k \to 0$ to 0 at $k \to \infty$ is monotonic since the derivative is positive at all k. The intersection point of this mixed mode with the spurious anti-Kelvin mode, $\omega_0 = -\alpha^{1/2}k$, occurs at $2\alpha^{1/2}k^2 = 1$; i.e., $k_{int} = 2^{-1/2}\alpha^{-1/4}$ at which wavelength the frequency is $(\omega_0)_{int} = -\left(2^{-1/2}\right)\alpha^{1/4}$. This intersection point is taken to be the point where the mixed mode changes its character from an Inertia-Gravity mode to Planetary mode (i.e., the point in Fig. 1.3 where the positive $n = 0$ mode changes from dashed to solid). However, as discussed above, the $\omega_0 = -\alpha^{1/2}k$ root is, in fact, a spurious solution of the dispersion relation on the unbounded plane due to the indefinite increase of $\eta(\phi)$. It is unclear whether the mixed mode exists as a continuous single mode in cases where the anti-Kelvin mode is either a genuine (and not spurious) mode or when it is not a solution of the dispersion curve at all (e.g., the narrow-channel asymptote or a sphere).

In summary, three conclusions can be drawn from the present channel theory. The first is that there exist Harmonic wave solutions in a narrow channel on the equatorial β-plane just as they exist in a narrow channel on the mid-latitude β-plane. The second conclusion is that the unbounded equatorial β-plane theory does not apply to a barotropic ocean since it is only relevant when $\delta\phi \geq 5\alpha^{1/4}$ which in a barotropic α is too wide for the first-order expansion, $\sin\phi \approx \phi$, to be valid. The third is that the mixed mode exists when the anti-Kelvin equatorial wave is a spurious solution of the dispersion relation that belongs to a singular eigenfunction. In contrast, when the anti-Kelvin mode is either a genuine solution or does not exist as a root of the dispersion relation, such as in a channel or on a sphere the mixed mode disappears. In the latter cases, the intersection between these two modes at k_{int} (which implies that the two modes coalesce) has to be carefully examined by taking into account the unstable nature of coalescence between two real waves over a certain wavelength range and given the uncertainty regarding the different nature of both the anti-Kelvin wave and the mixed mode at $k > k_{int}$ and $k < k_{int}$.

References

Erlick C, Paldor N, Ziv B (2007) Linear waves in a symmetric equatorial channel. Quart J Roy Met Soc 133(624):571–577. doi:10.1002/qj.44
Matsuno T (1966) Quasi-geostrophic motion in the equatorial area. J Meteorol Soc Jpn 44:25–43
Schiff LI (1968) Quantum mechanics. McGraw-Hill, New York, p 544

Chapter 4
Planetary and Inertia-Gravity Waves in an Equatorial Channel on a Sphere

4.1 Introduction

Some of the issues encountered in the preceding chapter in the development of wave theory in a channel on the equatorial β-plane might be resolved when the β-plane approximation is relaxed and the same problem is studied in an equatorial channel on a sphere. This problem will not suffer from the limitation associated with the linear expansion of Coriolis frequency, $\sin \phi$, but the price we should expect to pay for the more general approach is the much more complex form of the shallow water equations in spherical geometry in which the GRAD and DIV differential operators contain latitude-dependent coefficients. Another reason for studying the problem of an equatorial channel on a sphere is that this setup is an intermediate step that should be studied prior to studying the shallow water equations on the entire rotating sphere.

As in the previous chapters, the dimensional system (1.9) is non-dimensionalized by scaling the $u, v, \eta,$ and t variables on the same scales as in Chap. 2: $u, v,$ on $2\Omega a; t$ on $(2\Omega)^{-1}$; and η on H. The latitude and longitudes angles, ϕ and λ, are pure numbers when expressed in Radians, so they need not be scaled. With these scales, the non-dimensional form of system (1.9) in spherical coordinates is:

$$
\begin{aligned}
\frac{\partial u}{\partial t} - v \sin \phi &= -\frac{\alpha}{\cos \phi}\frac{\partial \eta}{\partial \lambda}, \\
\frac{\partial v}{\partial t} + u \sin \phi &= -\alpha \frac{\partial \eta}{\partial \phi}, \\
\frac{\partial \eta}{\partial t} &= -\frac{1}{\cos \phi}\left(\frac{\partial u}{\partial \lambda} + \frac{\partial (v \cos \phi)}{\partial \phi}\right).
\end{aligned}
\tag{4.1}
$$

As in the equations derived in the preceding chapters, the only non-dimensional parameter in these equations is $\alpha = gH/(2\Omega a)^{1/2}$ and as in all zonal channel

© The Author(s) 2015   35
N. Paldor, *Shallow Water Waves on the Rotating Earth*,
SpringerBriefs in Earth System Sciences, DOI 10.1007/978-3-319-20261-7_4

problems the boundary conditions imposed on the solutions of this system are the vanishing of the meridional velocity component at the channel walls, i.e., $v(\phi = \pm^{1}/_{2}\delta\phi) = 0$ where $\delta\phi$ is the channel width. Letting the solution of system (4.1) vary as $e^{i(k\lambda - Ct)}$ (so the wave's frequency ω equals kC) yields the following system for the ϕ-dependent amplitudes of u, V and η:

$$
\begin{bmatrix}
0 & \sin\phi & \frac{\alpha}{\cos\phi} \\
\frac{\sin\phi}{k^2} & 0 & \frac{\alpha}{k^2}\frac{\partial}{\partial\phi} \\
\frac{1}{\cos\phi} & \tan\phi - \frac{\partial}{\partial\phi} & 0
\end{bmatrix}
\begin{bmatrix} u \\ V \\ \eta \end{bmatrix}
= C
\begin{bmatrix} u \\ V \\ \eta \end{bmatrix},
\tag{4.2}
$$

where $V = iv/k$. Due to the inherent 2π periodicity of λ in spherical geometry the admissible values of k must be integers. The matrixlike eigenvalue problem (4.2) that contains ϕ-derivatives and not just functions or numbers is the spherical geometry counterpart of the Cartesian system (2.2). Similar to (2.2) the eigensolutions of (4.2) can be calculated numerically using, for example, a Chebyshev collocation method to compute the eigenvalue C and the eigenvector $(u(\phi), V(\phi), \mu(\phi))$ subject to the boundary conditions $V(\phi = \pm\delta\phi/2) = 0$.

Prior to presenting the numerical solutions of (4.2), it is instructive to draw some analytic conclusions based on a transformation of this system to a second-order eigenvalue equations similar to (2.6). The u-momentum equation (the first line in (4.2)) can be inverted to express u as a linear combination of V and η:

$$
u = \frac{V\sin\phi + \eta\frac{\alpha}{\cos\phi}}{C}.
\tag{4.3}
$$

This relationship can be used to eliminate u from the other two equations in (4.2) to obtain the following second-order system in V and η which is the spherical geometry counterpart of (2.4):

$$
\frac{\partial}{\partial\phi}
\begin{bmatrix} V\cos\phi \\ \eta \end{bmatrix}
= \frac{1}{C\cos\phi}
\begin{bmatrix}
\sin\phi & (\alpha - C^2\cos^2\phi) \\
\left(\frac{\omega^2 - \sin^2\phi}{\alpha}\right) & -\sin\phi
\end{bmatrix}
\begin{bmatrix} V\cos\phi \\ \eta \end{bmatrix}.
\tag{4.4}
$$

This system differs from its planar counterpart substantially as there are many more ϕ-dependent terms in (4.4) than y-dependent terms in (2.4). The reason for this difference results directly from the inherent complexity of spherical coordinates where the (dimensional) length Δx associated with a fixed longitude angle, $\Delta\lambda$, is latitude-dependent $\Delta x = a\cos\phi\Delta\lambda$ which implies $\partial/\partial\lambda = (a\cos\phi)^{-1}\partial/\partial x$. The non-dimensional form of this relation is $\partial/\partial\lambda = (\cos\phi)^{-1}\partial/\partial x$, so the non-dimensional wave number k and phase speed C of Cartesian coordinates transform to $k/\cos\phi$ and $C\cos\phi$ in spherical coordinates while the phase $k_{\text{plane}}x$ equals the phase $k_{\text{sphere}}\lambda$ (so the frequency $\omega = kC$ is unchanged by the change of coordinates). Another difference between the two systems is the change from V as the dependent variables in (2.4) to $V\cos\phi$ in (4.4) which results from the presence of $\cos\phi$ coefficients in the non-dimensional meridional component of the divergence operator (this can be appreciated by considering a uniform meridional velocity, V, for which DIV(V) is zero on a plane, while on a sphere it equals $V\tan\phi$). The last difference between (2.4) and (4.4) results from the expansion of the non-dimensional Coriolis parameter $f = \sin\phi \approx \sin\phi_0 + y\cos\phi_0$ in Cartesian coordinates while in the spherical system, (4.4), f appears in its exact form.

A similar transformation to that used in transforming (2.4)–(2.6) can also be applied to (4.4) but the singular $C^2 = \alpha$ case that yields Kelvin waves in Cartesian coordinates is more complex in spherical coordinates since setting in (4.4) $C^2 = \alpha$ only implies that the coefficient of η in the equation for $V\cos\phi$ vanishes at particular latitude in the channel but the equations remain coupled so the $V\cos\phi$ equation cannot be solved independently of the η-equation as in Sect. 2.1.

Despite the difference between Cartesian and spherical coordinates associated with the singular nature of Kelvin waves the elimination of η from (4.4) to obtain a second-order equation for $V\cos\phi$ can be done straightforwardly by following the same procedure employed in Chap. 2. Accordingly, differentiating the equation for $V\cos\phi$ with respect to ϕ and substituting the expressions for η and $\partial\eta/\partial\phi$ from the two first-order equations yields the following second-order equation for $V\cos\phi$:

$$\frac{\partial^2(V\cos\phi)}{\partial\phi^2} - \left(\tan\phi\frac{\alpha + C^2\cos^2\phi}{\alpha - C^2\cos^2\phi}\right)\frac{\partial(V\cos\phi)}{\partial\phi}$$
$$+ \left[\frac{\omega^2}{\alpha} - \frac{1}{C} - \frac{\sin^2\phi}{\alpha} - \frac{k^2}{\cos^2\phi} + \tan^2\phi\left(\frac{\alpha + 2C\cos^2\phi}{\alpha - C^2\cos^2\phi}\right)\right]V\cos\phi = 0. \tag{4.5}$$

Substituting $\psi(\phi)$, defined by,

$$\psi(\phi) = V(\phi)\cos\phi\left(\frac{\alpha}{C\cos\phi} - C\cos\phi\right)^{-1/2} \tag{4.6}$$

for $V\cos\phi$ transforms Eq. (4.5) to the Schrödinger-like equation:

$$\alpha\frac{\partial^2\psi}{\partial\phi^2} + [E - U_1(\phi) - \alpha U_2(\phi)]\psi = 0, \tag{4.7}$$

where

$$E = k^2C^2 - \frac{\alpha}{C} = \omega^2 - \frac{\alpha}{C},$$
$$U_1(\phi) = \sin^2\phi + \frac{\alpha k^2}{\cos^2\phi},$$
$$U_2(\phi) = \frac{3}{4}\tan^2\phi\left(\frac{\alpha + C^2\cos^2\phi}{\alpha - C^2\cos^2\phi}\right)^2 - \frac{1}{2}\left(\frac{\alpha + C^2\cos^2\phi}{\alpha - C^2\cos^2\phi}\right)$$
$$- \tan^2\phi\left(\frac{\alpha + 2C\cos^2\phi}{\alpha - C^2\cos^2\phi}\right). \tag{4.8}$$

As in the planar cases studied in the previous chapters E is the energy of the Schrödinger-like Eq. (4.7) and according to its definition in (4.8) every energy value, E_n, is associated with three phase speeds given by the roots of the cubic: $k^2 C^3 - E_n C - \alpha = 0$ (or in terms of frequency: $\omega^3 - E_n \omega - k\alpha = 0$). The potential of the Schrödinger-like Eq. (4.7) is $U_1(\phi) + \alpha U_2(\phi)$. The boundary conditions that solutions of Eq. (4.7) are required to satisfy are the vanishing of the meridional velocity $v = -ikV$ at the channel walls located at $\phi = \pm\delta\phi/2$, i.e., $\psi(\phi = \pm\delta\phi/2) = 0$ according to (4.6).

When the spherical eigenvalue Eq. (4.7) with the $\psi(\phi = \pm\delta\phi/2) = 0$ boundary conditions is compared to the planar problem (2.6) is becomes evident that the differences between two equations described above (following the derivation of the $(V \cos \phi, \eta)$ system (4.4)) explain the changes in E and $U_1(\phi)$ in (4.8) compared to (2.6) or (3.3) while $U_2(\phi)$ consists of terms that have no counterparts on the β-plane(s). The terms in $U_2(\phi)$ originate from the terms of (4.4) that include $C \cos \phi$ coefficients and their derivatives. These "purely spherical" terms are the source of the extra terms in (4.5) including the first derivative term $\partial(V \cos \phi)/\partial\phi$ that have no counterpart in (2.6).

The spherical geometry is expected to modify the dispersion relations and eigenfunctions of the planar waves but not to generate new types of waves because from a physical perspective each wave type originates from a different balance of forces and these forces exist, perhaps in slightly different forms, in the two geometries. Thus, the "purely spherical" potential $U_2(\phi)$ can be expected to alter the planar theory only marginally. A more formal justification for the neglect of $U_2(\phi)$ is that it is multiplied by α which on earth is smaller than 1, ranging from 5×10^{-6} in a baroclinic ocean to 5×10^{-2} in a barotropic ocean, so the full potential in (4.7), $U_1(\phi) + \alpha U_2(\phi)$, can be safely approximated by $U_1(\phi)$. The neglect of $\alpha U_2(\phi)$ is uniformly valid when $C^2 < \alpha$ (i.e., Planetary waves) since in this case the denominator of the three $U_2(\phi)$ terms: $\alpha - C^2 \cos^2 \phi$ is definite positive at all ϕ (i.e., it never vanishes) so $\alpha U_2(\phi)$ can be safely neglected compared to $U_1(\phi)$. In the case of the fast-moving Inertia-Gravity (Poincaré) waves with $C^2 > \alpha$, the denominator of the three $U_2(\phi)$ terms, $\alpha - C^2 \cos^2 \phi$, vanishes only at high enough latitudes where the eigenfunction $\psi(\phi)$ is zero. Thus, in the context of the present equatorial channel problem where the domain is bounded by $\pm\delta\phi/2$ the singularity of $U_2(\phi)$ at sufficiently high latitudes does not seriously limit the applicability of the results. These heuristic arguments have to be substantiated quantitatively to validate the approach taken below in developing the spherical theory by neglecting $\alpha U_2(\phi)$.

The approximate system obtained by neglecting $\alpha U_1(\phi)$ in (4.7) is the Schrödinger equation:

$$\frac{\partial^2 \psi}{\partial \phi^2} + \left[\frac{E}{\alpha} - \frac{U_1(\phi)}{\alpha}\right]\psi = 0, \tag{4.9}$$

where E and $U_1(\phi)$ are defined in (4.8).

Given the symmetry of $U_1(\phi)$ and the bounded ϕ-domain, $|\phi| \leq \delta\phi/2$, an expansion of $U_1(\phi)$ to second order in ϕ can be expected to yield estimates that are correct to order ϕ^4. The resulting eigenvalue problem that approximates (4.9) then becomes:

$$\frac{\partial^2 \psi}{\partial \phi^2} + \left[\frac{E}{\alpha} - k^2 - \left(\frac{1 + \alpha k^2}{\alpha}\right)\phi^2\right]\psi = 0; \quad \psi\left(\phi = \pm\frac{\delta\phi}{2}\right) = 0. \qquad (4.10)$$

Figure 4.1, adaped from De-Leon et al. (2010), compares several thousand solutions of this approximate system with those of the exact system (4.2) for two $\delta\phi/2$ values (0.05 and 0.4 rad); four α values ranging from 10^{-6} to 10^{-3}; and several hundred values of k for the three phase speeds. These results clearly show that for these baroclinic α values, the neglect of $\alpha U_2(\phi)$ as well as the expansion of $U_1(\phi)$ to second order in ϕ entail minute changes to the numerically calculated phase speeds and the combination $k^2 C^2/\alpha - 1/C$.

Having established numerically the accuracy of the approximate eigenvalue Eq. (4.10), we can now turn to its analytic inferences. The quadratic form of the approximate potential suggests that a similar transformation to that used in deriving the eigenvalue equation on the equatorial β-plane, (3.3), can also be used here.

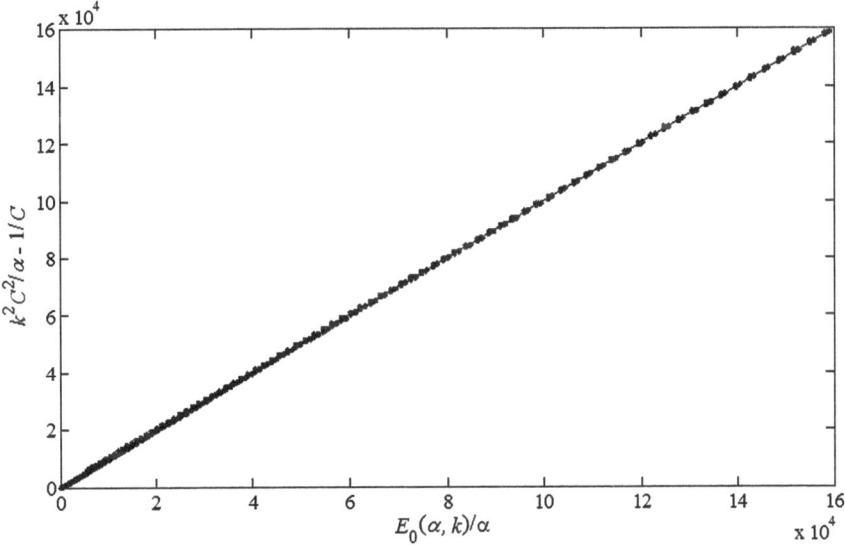

Fig. 4.1 Values of $k^2 C^2/\alpha - 1/C$ obtained from numerical solutions of (4.2) versus values of $E_0(k, \alpha)/\alpha$ obtained from numerical solutions of the approximate system (4.10). Several thousand points are shown in this plot in which the C values belong to Rossby waves and the two Poincaré waves. Data include 4 α-values ranging from 10^{-6} to 10^{-3} and k-values ranging from 1 to several hundred. Channel widths, $\delta\phi/2$, are 0.05 and 0.4. Permission from Co-Action/TellusA: doi:10.1111/j.1600-0870.2009.00420.x

Toward that end we transform ϕ, the independent variable of (4.10), to the independent variable: $z = \phi/(\delta\phi/2)$ in terms of which (4.10) is written as:

$$\frac{\partial^2 \psi}{\partial z^2} + \left[\left(\frac{\delta\phi}{2}\right)^2 \left(\frac{E}{\alpha} - k^2\right) - \left(\frac{\delta\phi}{2}\right)^4 \left(\frac{1 + \alpha k^2}{\alpha}\right) z^2\right] \psi = 0; \quad \psi(z = \pm 1) = 0.$$

$$(4.11)$$

This eigenvalue problem is the spherical counterpart of the planar problem (3.3) and when the constants that appear in the two problems are compared it becomes evident that the eigenvalues $(\omega^2/\alpha - 1/C - k^2) \times (\delta\phi/2)^2$ are identical in the two problems, while the coefficient of z^2 in the planar problem studied in (3.3), $(\delta\phi/2)^4(1/\alpha)$, is modified in the spherical problem to $(\delta\phi/2)^4(1/\alpha + k^2)$. This additional k^2 term in the coefficient of the potential originates from the latitude dependence of the zonal wave number on a sphere as explained in the discussion of the difference between (4.4) and (2.4). In addition to this difference in the coefficient of z^2 in the potentials of the two problems the two problems also differ in the dependent variable which is V in the planar problem and Ψ (related to $V \cos \phi$ by (4.6)) in the spherical problem.

Aside from the somewhat different expressions of the coefficient of z^2 in (4.11) compared to that of (3.3) the solutions are expected to be identical in the two problems. Since the potential vanishes for narrow channels the asymptotic solutions of (4.11) are identical to (3.5), i.e.,

$$\psi_n(z) = A \sin\left(\frac{(n + 1)\pi}{2}(z + 1)\right), \quad n = 0, 1, 2, \ldots \qquad (4.12)$$

and the associated eigenvalues are given by (3.6) so the energy levels of (4.8) can be derived from the solutions of (3.2) obtained in Chap. 3:

$$E_n = k^2 C_n^2 - \frac{1}{C_n} = \alpha\left(k^2 + \left(\frac{(n + 1)\pi}{\delta\phi}\right)^2\right), \quad n = 0, 1, 2, \ldots \qquad (4.13)$$

The wide-channel asymptotic solution of (4.11) is obtained from (3.7) and (3.8) by substituting $(1/\alpha + k^2)^{1/2}$ for $\alpha^{-1/2}$ in the coefficient of z^2 in (3.3) (in which b/ε is given by (3.4)) which yields the wide-channel asymptotic solutions on a sphere:

$$\psi_n(z) = A \cdot H_n\left(\frac{\delta\phi}{2}\left(\frac{1}{\alpha} + k^2\right)^{1/4} z\right) \cdot e^{-\left(\frac{\delta\phi}{2}\right)^2 \frac{z^2}{2}\sqrt{\frac{1}{\alpha} + k^2}}, \quad n = 0, 1, 2, \ldots \qquad (4.14)$$

The associated eigenvalues are found from the known relation between the eigenvalues and the coefficient of z^2 in the Harmonic Oscillator of Quantum Mechanics:

$$\left(\frac{\delta\phi}{2}\right)^2 \left(\frac{E_n}{\alpha} - k^2\right) = (2n+1)\left(\left(\frac{\delta\phi}{2}\right)^2 \left(\frac{1}{\alpha} + k^2\right)^{1/2}\right), \quad n = 0, 1, 2, \ldots$$

Substituting the definition of E from (4.8) and rearrangement yields:

$$\frac{\omega_n^2}{\alpha} - \frac{k}{\omega_n} - k^2 = (2n+1)\left(\frac{1}{\alpha} + k^2\right)^{1/2}, \quad n = 0, 1, 2, \ldots \qquad (4.15)$$

This cubic relation is the spherical counterpart of (3.11), the wide-channel relation on the β-plane.

Figure 4.2, adapted from De-Leon et al. (2010), shows the transition in the structure of the $V(\phi)$ eigenfunctions of the $n = 0$ mode from harmonic in the narrow-channel asymptote, (4.12), to Gaussian in wide-channel asymptote (4.14). In this figure α is varied while $\delta\phi/2$ is held fixed at 0.2 but a similar transition occurs for fixed α when $\delta\phi/2$ in varied.

Given the similarity between the eigensolutions derived in the problem on a sphere and those derived in Chap. 3 on the β-plane the only point that requires further consideration is the $n = 0$ modes that were derived in Chap. 3 as a particular case where the cubic is decomposed into an exact form that substitutes the approximate forms based on high/low frequency.

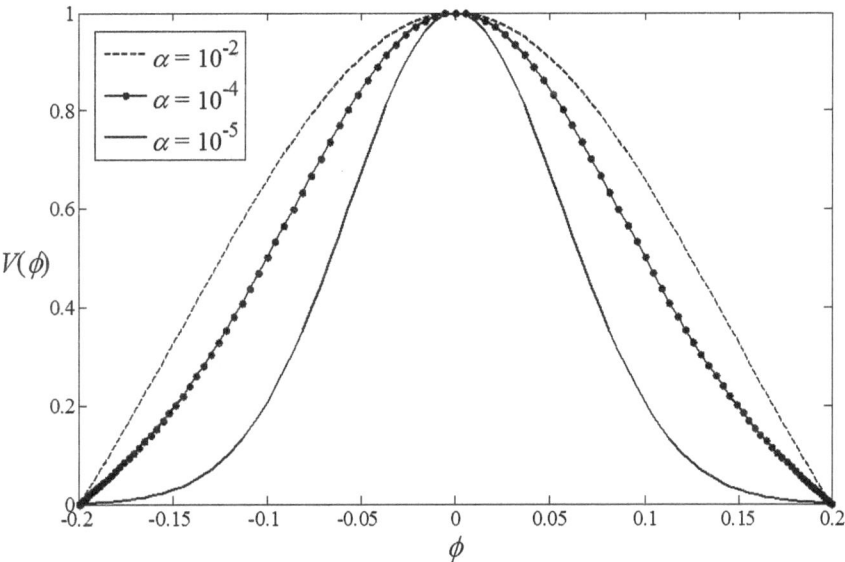

Fig. 4.2 The transition of $V(\phi)$ from harmonic to Gaussian when α is decreased from 10^{-2} (*narrow-channel asymptote*) to 10^{-5} (*wide-channel asymptote*) when $\delta\phi/2 = 0.2$. Permission from Co-Action/TellusA: doi:10.1111/j.1600-0870.2009.00420.x

As is the case of the β-plane, on a sphere too, the roots of (4.15) can be characterized based on the magnitude of ω_0 but a decomposition of the cubic (4.15) with $n = 0$ to the spherical counterpart of (3.14) is impossible. The reason for this is that while the RHS of (4.15) depends on k, the RHS of (3.11) is independent of it and this independence on k enables the exact decomposition of the latter that yields (3.14) for $n = 0$. To examine the possibility that two of the three frequencies of (4.15) intersect for some combination of k, n and α, we multiply the equation through by $\omega_n \alpha^{-1/2} k^{-3}$ which yields the cubic:

$$X^3 - X - X\left[(2n+1)\left(F^2 + \alpha^{1/2}F\right)^{1/2}\right] - F = 0, \tag{4.16}$$

where $X = \frac{\omega_n}{k\alpha^{1/2}}$ and $F = 1/(\alpha^{1/2}k^2)$. Since $\frac{X^3-X}{X+1} = X^2 - X$, $X + 1$ is a root of (4.16), i.e., $\omega_0 = -\alpha^{1/2}k$ is a root of (4.15), if and only if $(2n+1)(F^2 + \alpha^{1/2}F)^{1/2} = F$, and when the square of this condition is divided through by $F \neq 0$ one obtains $4n(n+1)F = -(2n+1)^2\alpha^{1/2}$ which can be satisfied only when the two sides equal 0, i.e., when $n = 0$ and $\alpha = 0$. The origin of the $\alpha^{1/2}F$ term in (4.16) can be traced back to the αk^2 term in the coefficient of z^2 in (4.11), i.e., the $\alpha k^2/\cos^2 \phi$ term in the potential $U_1(\phi)$ in (4.8). In contrast, as is evident from (3.4), in the planar problem k does not affect the potential and it appears only in the expression of energy. Since the latitude dependence of $k^2/\cos^2\phi$ is inherent to the spherical geometry and is the very basic difference between Cartesian and spherical coordinates the existence of the anti-Kelvin mode in the dispersion relation seems to be unique to Cartesian coordinates.

Though it is not possible to derive explicit approximate expressions for the three roots of (4.16) it is possible to examine whether the two negative roots intersect one another for certain values of the coefficients of this cubic. For such an intersection to take place the frequency of the negative Inertia-Gravity wave must equal that of the Planetary wave at some wave number k. Since the sum of the three roots of any cubic equals the negative of the coefficient of X^2, which is 0 in (4.16), the two intersecting negative roots of (4.16) should be identical in their absolute values and each of them equals half the positive root. Thus, an intersection of the two negative roots implies that the cubic (4.16) has the form $(X - 2X_1)(X + X_1)^2 = 0$ where X_1 is the absolute value of the negative double root. Equating the coefficients of X^0 and X^1 in the two forms yields the relations $2X_1^3 = F$ and $3X_1^2 = 1 + (2n+1)(F^2 + \alpha^{1/2}F)^{1/2}$, respectively. Since both relations have to be satisfied for the two negative roots of (4.16) to intersect one obtains:

$$3\left(\frac{F}{2}\right)^{2/3} - (2n+1)\left(F^2 + \alpha^{1/2}F\right)^{1/2} = 1. \tag{4.17}$$

In the planar case studied in Chap. 3 an intersection of modes occurs for $n = 0$. In contrast, the calculations of the two sides of (4.17) shown in Fig. 4.3 demonstrate

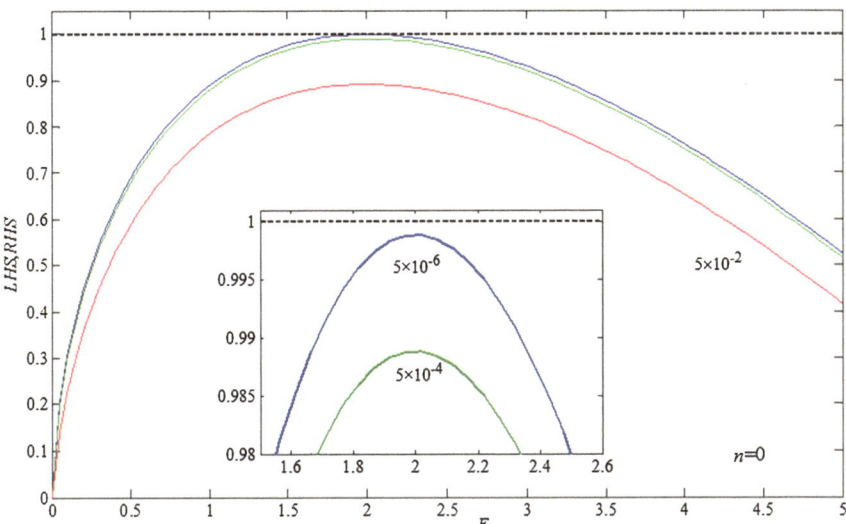

Fig. 4.3 The LHS of (4.17) as a function of F for the indicated values of α and for $n = 0$ (*solid lines*). The horizontal *dashed line* is "1", the RHS of (4.17), that has to be intersected by the LHS curves to satisfy the mode-crossing condition. The *inset* is a zoom in on the region where the $\alpha = 5 \times 10^{-6}$ curve nearly intersects "1" but does not

that the intersection condition, (4.17) with $n = 0$, is not satisfied for any acceptable nonzero value of α. Thus, on a sphere, no mode crossing occurs even for $n = 0$ and even for baroclinic α value of 5×10^{-6}.

The dispersion diagrams, $\omega(k)$, of the three waves given in (4.15) are shown in Fig. 4.4 for $\alpha = 5 \times 10^{-6}$ and in Fig. 4.5 for $\alpha = 5 \times 10^{-2}$ both for $n = 0,1,\ldots5$. For the baroclinic α in Fig. 4.4 the $n = 0$ Rossby mode and the $n = 0$ Poincarè mode nearly intersect near $k = 15$, but the zoom-in panel shows that a gap exists between these modes. As anticipated by the results of Fig. 4.3, the gap between the two modes widens with the increase in α and at $\alpha = 5 \times 10^{-2}$ the separation between the two modes is clearly discernible even in the large-scale plot of Fig. 4.5.

The near, but not actual, crossing of the Inertia-Gravity and Planetary $n = 0$ modes on a sphere shown in Figs. 4.4 and 4.5 provides a new perspective of the emergence of the mixed mode, as well as the disappearance of the anti-Kelvin mode, on the unbounded equatorial β-plane (Fig. 1.2). The anti-Kelvin mode is omitted from the dispersion relation based on the singular behavior of its associated eigenfunctions in the limit $y \rightarrow \infty$. However, the existence of this mode in the dispersion curves determines the point where the mixed mode transforms from Inertia-Gravity at small k to Planetary at large k. In the planar approximation k appears only in the energy and not in the potential (i.e., the planar counterpart of the potential in Eq. (4.10) is ϕ^2/α which is independent of k). This independence of the planar potential on k brings about the intersection between the $n = 0$ Inertia-Gravity and Planetary modes that forms the mixed mode even when the

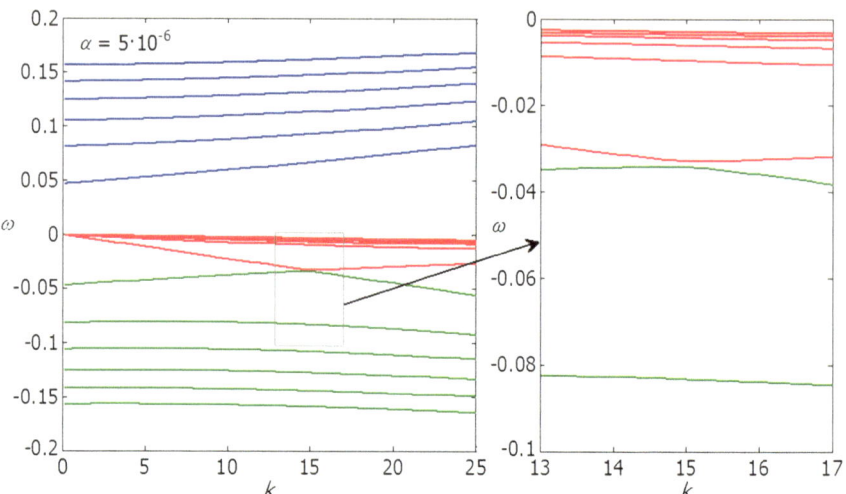

Fig. 4.4 The dispersion relations of the three waves of (4.15) for $\alpha = 5 \times 10^{-6}$ and $n = 0, 1…5$. The *right panel* is a zoom in on the region of near intersection between the two $n = 0$ negative modes and this zoom in clarifies that the dispersion curves of the two modes do not intersect

anti-Kelvin mode is eliminated from the dispersion relation (so the mixed mode and the anti-Kelvin mode do not intersect). In contrast, the spherical potential depends on k so the anti-Kelvin mode does not appear at all in the dispersion relation and each of the two negative $n = 0$ modes includes a small-k and a large-k segments.

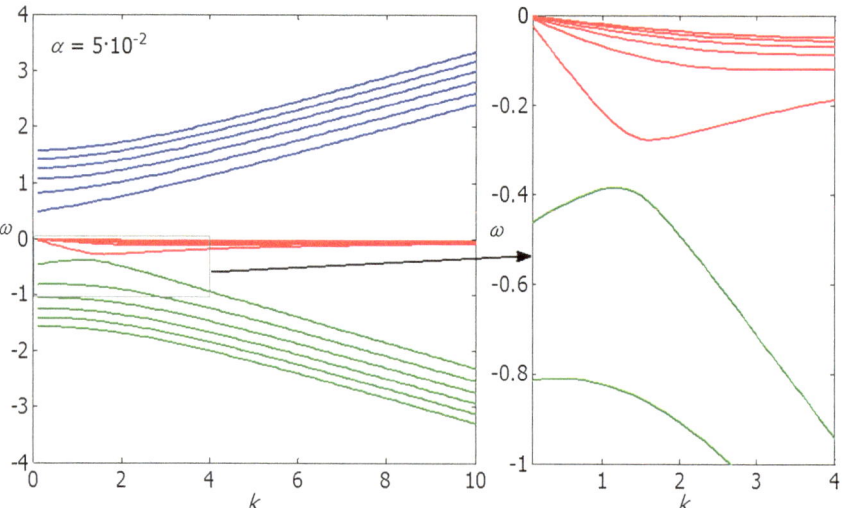

Fig. 4.5 As in Fig. 4.4 but for $\alpha = 5 \times 10^{-2}$. The *right panel* is a zoom-in on the region where the frequencies of the two $n = 0$ negative modes approach one another

The finding outlined above regarding the disappearance of both the mixed mode and the anti-Kelvin wave from the dispersion relation on a sphere suggests that the presence of these modes in the dispersion relation on the unbounded β-plane might result from the overly simplified setup of this theory where $|y| \to \infty$ despite the first-order expansion of the Coriolis frequency.

Reference

De-Leon Y, Erlick C, Paldor N (2010) The eigenvalue equations of equatorial waves on a sphere. Tellus A 62A:62–70

Chapter 5
Planetary and Inertia-Gravity Waves in a Mid-latitude Channel on a Sphere

Having examined in the previous chapter the relationship between waves in an equatorial channel on the β-plane and on a sphere it is only natural to do the same for a mid-latitude channel. Though mid-latitude waves do not include a unique wave such as the equatorial mixed mode it is still unclear whether or how the inclusion of several latitude-dependent terms associated with the spherical geometry will modify the planar solution derived in Chap. 2 in which only the Coriolis frequency, $f(y)$, was assumed to vary with latitude.

The equations derived in Chap. 4 for zonally propagating waves in spherical coordinates apply also to the problem of a mid-latitude channel on a sphere addressed in this chapter. The only difference between the two problems is in the application of the $V = 0$ boundary conditions, which are applied in this chapter at ϕ_s and ϕ_n, the equatorward and poleward channel walls, respectively. In contrast to the different boundary conditions in the two problems the exact differential equation, Eq. (4.5) (or its counterpart, Eq. (4.7)), the approximate differential equation, Eq. (4.9), the definition of ψ, Eq. (4.6) and the definitions of E and U_1 in Eq. (4.8) are all relevant without any change to the present mid-latitude problem. Accordingly, the approximate eigenvalue problem (equations and boundary conditions) to be solved is:

$$\alpha \frac{\partial^2 \psi}{\partial \phi^2} + \left[E - \left(\sin^2 \phi + \frac{\alpha k^2}{\cos^2 \phi} \right) \right] \psi = 0, \quad \psi(\phi_s) = 0 = \psi(\phi_n) \qquad (5.1)$$

where ϕ_s and ϕ_n are both positive (negative) in the Northern (Southern) Hemisphere and where:

$$\psi(\phi) = V(\phi) \cos \phi \left(\frac{\alpha}{C \cos \phi} - C \cos \phi \right)^{-1/2} \quad \text{and} \quad E = k^2 C^2 - \frac{\alpha}{C} = \omega^2 - \frac{\alpha}{C}. \qquad (5.2)$$

Note that the spherical eigenvalue problem (5.1) does not reduce to the planar problem (2.7) when $\sin \phi$ is expanded to first order in $\phi - \phi_0$ where ϕ_0 is, say, the channel's midpoint $(\phi_s + \phi_n)/2$ due to the ϕ dependence of the term $k^2 / \cos^2 \phi$ in

© The Author(s) 2015
N. Paldor, *Shallow Water Waves on the Rotating Earth*,
SpringerBriefs in Earth System Sciences, DOI 10.1007/978-3-319-20261-7_5

spherical coordinates (while the counterpart of this term on a plane is the constant wavenumber k, incorporated in the expression for the energy). The potential of (5.1), $\sin^2 \phi + \alpha k^2 / \cos^2 \phi$, can be expanded to first order in $\phi - \phi_s$ as was done in the planar problem studied in Chap. 2. Carrying out this expansion in (5.1) and dividing it through by α yields the spherical counterpart of (2.14):

$$\frac{\partial^2 \psi}{\partial \phi^2} + \left[\frac{E}{\alpha} - \frac{\sin^2 \phi_s}{\alpha} - \frac{k^2}{\cos^2 \phi_s} - 2 \tan \phi_s \left(\frac{\cos^2 \phi_s}{\alpha} + \frac{k^2}{\cos^2 \phi_s} \right) (\phi - \phi_s) \right] \psi = 0.$$

(5.3)

Defining

$$Z = - \left(\frac{E}{\alpha} - \frac{\sin^2 \phi_s}{\alpha} - \frac{k^2}{\cos^2 \phi_s} \right) b^{-2/3} + (\phi - \phi_s) b^{1/3} \qquad (5.4)$$

where $b = 2 \tan \phi_s \left(\frac{\cos^2 \phi_s}{\alpha} + \frac{k^2}{\cos^2 \phi_s} \right)$ and substituting Z for $\phi - \phi_s$ in Eq. (5.3) transforms this equation into Airy equation:

$$\frac{\partial^2 \psi}{\partial Z^2} - Z \psi = 0.$$

As in Chap. 2 (see the paragraph following (2.16)) the boundary condition $\psi(\phi_s) = 0$ is satisfied by setting the equatorward boundary, $\phi - \phi_s$, to be the nth zero of Ai(Z). Letting $Z(\phi = \phi_s) = -\xi_n$ (where ξ_n is the absolute value of the nth zero of Ai(Z)) and $E = E_n$ in (5.4) yields:

$$\frac{E_n}{\alpha} = \frac{\sin^2 \phi_s}{\alpha} + \frac{k^2}{\cos^2 \phi_s} + \left(2 \tan \phi_s \left(\frac{\cos^2 \phi_s}{\alpha} + \frac{k^2}{\cos^2 \phi_s} \right) \right)^{2/3} \xi_n. \qquad (5.5)$$

The second boundary condition, $\psi(\phi_n) = 0$, is satisfied, provided that ϕ_n is located at sufficiently large and positive values of Z such that the contribution of Bi(Z) to the solution can be neglected, while Ai(Z) itself can be considered zero at that point. Letting $Z(\phi = \phi_n) = 2$ and $E = E_n$ in (5.4) yields the following threshold value of channel width above which the Trapped wave theory is valid:

$$\phi_n - \phi_s \geq (2 + \xi_n) b^{-1/3} = \frac{(2 + \xi_n) \alpha^{1/3}}{\left(2 \tan \phi_s \left(\cos^2 \phi_s + \frac{\alpha k^2}{\cos^2 \phi_s} \right) \right)^{1/3}}. \qquad (5.6)$$

For $\alpha k^2 \ll \cos^4 \phi_s$ this bound reduces to its planar counterpart, (2.19). As in the planar problem, in narrow channels, where (5.6) is violated, harmonic solutions of (5.1) are obtained by evaluating the potential in mid-channel, i.e., at $\phi_0 = (\phi_s + \phi_n)/2$ in which case the equation becomes a constant coefficient equation.

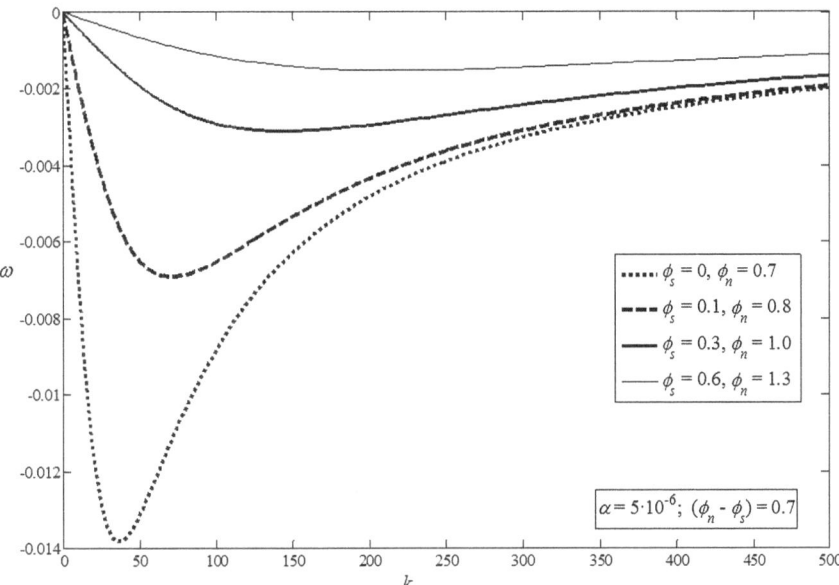

Fig. 5.1 Numerical calculation of the dispersion relation of Rossby waves based on the exact eigenvalue system (4.2) subject to the boundary conditions $V(\phi_n) = 0 = V(\phi_s)$. The value of ϕ_s has a strong effect on the dispersion relations. Permission from American Meteorological Society: J. Phys. Oceanogr. doi:10.1175/2009JPO4083.1

The analytic insight gained by analyzing the approximate Schrödinger equation (5.1) must be validated by comparing it to numerical solutions of the exact eigenvalue system, (4.2), subject to the boundary conditions $V(\phi_n) = 0 = V(\phi_s)$. The calculations of the dispersion relations of Rossby waves shown in Fig. 5.1 for $\alpha = 5 \times 10^{-6}$ and a fixed channel width of 0.7 rad show that the location of the southern boundary (indicated in the insets of the figure in Radians) affects the dispersion relation quite significantly (all the figures in this chapter were adapted from De-Leon and Paldor (2009)).

In contrast, the results shown in Fig. 5.2 for the case in which the equatorward wall is held fixed at $\phi_s = 0.3$ but the latitude of the northern wall is increased (i.e., the channel width increases with ϕ_n) clearly show that the dispersion relation is unaffected by the increase in the channel width beyond a threshold width of about 0.3 rad for $\alpha = 10^{-3}$. These results that are obtained from numerical solutions of the exact system (4.2) confirm the validity of the bounds in (5.6), which were derived from the approximate equation (5.1).

The analytic estimate in (5.6) of the threshold value of $\phi_n - \phi_s$ above which the non-harmonic theory is valid also predicts a decrease in this threshold value when k is increased. Conversely, in a channel of fixed width, Trapped waves will be realized at large k and Harmonic waves at small k. These predictions are confirmed by the results shown in Fig. 5.3 for a channel width $\phi_n - \phi_s = 0.1$ rad and $\alpha = 0.1$:

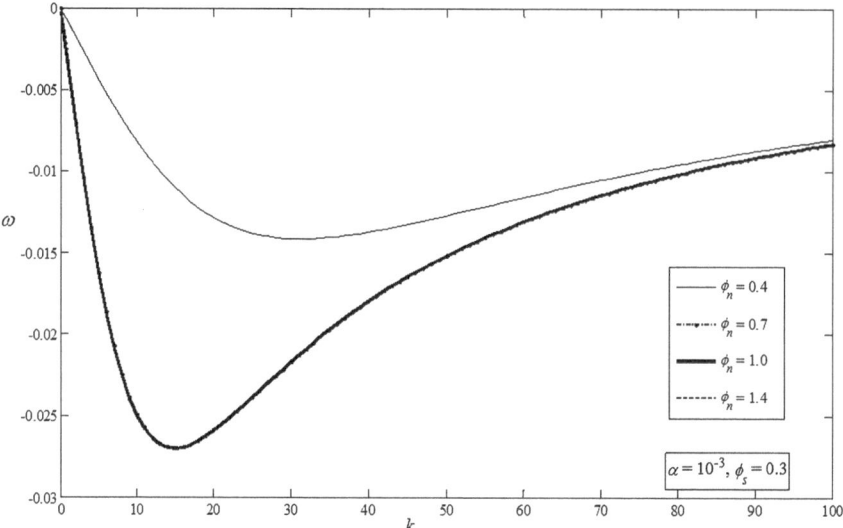

Fig. 5.2 The numerically calculated dispersion relation of the exact system (4.2) for the indicated values of ϕ_n and for $\phi_s = 0.3$. In agreement with the analysis of the approximate equation, (5.1), in wide channels with $\phi_n - \phi_s > 0.3$ rad, the width does not affect the dispersion relation. Permission from American Meteorological Society: J. Phys. Oceanogr. doi:10.1175/2009JPO4083.1

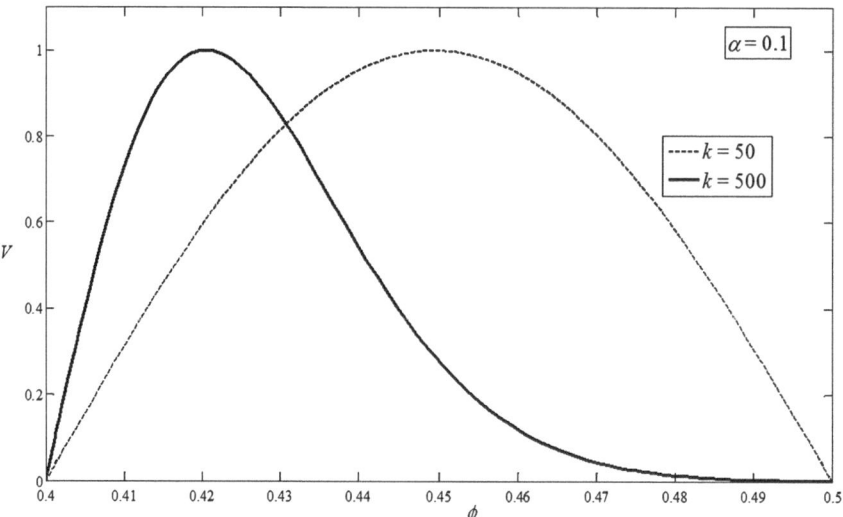

Fig. 5.3 The change in the meridional structure of the $V(\phi)$ eigenfunctions when k is increased from 50 to 500 in a narrow channel. Though the channel is only 0.1 rad wide, it becomes a wide channel for large k when αk^2 dominates the denominator of (5.6). Permission from American Meteorological Society: J. Phys. Oceanogr. doi:10.1175/2009JPO4083.1

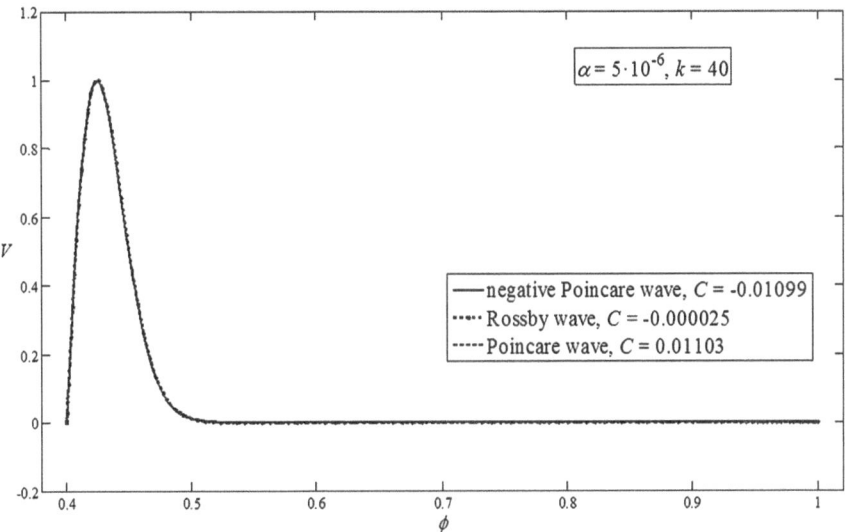

Fig. 5.4 The $V(\phi)$ eigenfunctions corresponding to the three values of C of two Poincarè modes and the Rossby mode for the indicated values of α and k. The u and η eigenfunctions differ between these modes since their relation to $V(\phi)$ involves C. Permission from American Meteorological Society: J. Phys. Oceanogr. doi:10.1175/2009JPO4083.1

The transition from harmonic structure to trapped structure is clearly evident in this channel when k increases from 50 to 500; that is, αk^2 is increased by 100 (from 250 to 25,000) so the threshold value on the RHS of (5.6) decreases by $100^{1/3}$ to 4.64.

Furthermore, the eigenfunctions shown in Fig. 5.4, which were obtained from numerical solutions of the exact system (4.2), clearly show the trapping near the equatorward wall anticipated by the $\mathrm{Ai}(Z(\phi))$ solution as well as the identical meridional structure of $V(\phi)$ of the two Poincaré modes and the Rossby mode. This identical latitudinal structure is a natural outcome of the unified Schrödinger equation formulation (5.1) of $\psi(\phi)$ for the three waves that is transformed to $V(\phi)$ via (5.2). However, the identical latitudinal structure of $V(\phi)$ of the three waves cannot be inferred from the exact set (4.2) in which each eigenvalue C is associated with a different corresponding set of (u, V, η) eigenfunctions.

Having established the qualitative and quantitative relevance of the analytic approximate solutions to the exact numerical results it remains to examine the quantitative match between the dispersion relations $\omega(k)$ derived from the value of E_n, (5.5), following the definition of $C(E)$ given in (5.2) with the dispersion relations calculated numerically from the exact system (4.2). The results shown in Fig. 5.5 confirm the validity of the approximate solutions at least in the range of parameters where (5.6) is satisfied.

The three dispersion relations of Rossby waves in a zonal channel in mid-latitudes: the Harmonic wave relation, the Trapped wave relation on the β-plane and the Trapped wave relation on a sphere are compared in Fig. 5.6. It is

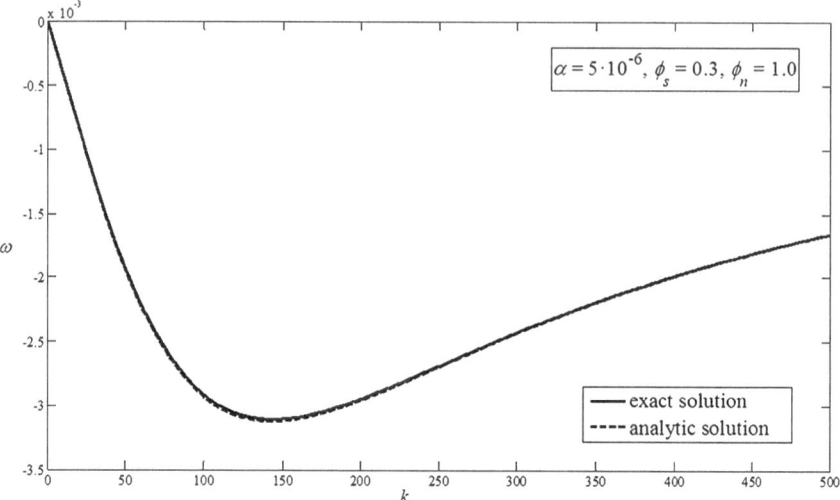

Fig. 5.5 A comparison between the dispersion relation of the first Rossby mode $n = 0$ obtained from exact numerical solution of (4.2) and the approximate analytical solutions. Permission from American Meteorological Society: J. Phys. Oceanogr. doi:10.1175/2009JPO4083.1

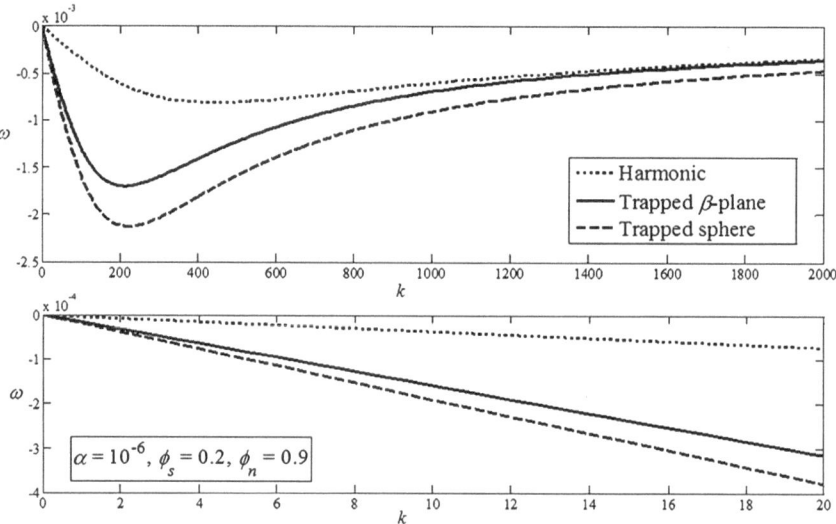

Fig. 5.6 The three dispersion relations of the $n = 0$ Rossby mode in a wide mid-latitude channel. The slopes of the $\omega(k)$ relations are the westward phase speed of the corresponding theory. The fastest wave is the Trapped wave on a sphere and the slowest one is the Harmonic planar wave. The *lower panel* is a zoom-in on the long-wave asymptote of the *top panel*. Permission from American Meteorological Society: J. Phys. Oceanogr. doi:10.1175/2009JPO4083.1

evident from this comparison that for the same channel width and value of α (i.e., 0.7 rad and 10^{-6}, respectively in the case of Fig. 5.6) the Harmonic wave theory yields the slowest westward propagation speed while the Trapped wave theory in the same channel on a sphere yields the fastest phase speed of the three. However, much of the difference between the spherical theory and the harmonic planar one is recovered by the planar Trapped wave theory.

Having completed the development of the Trapped wave theory in spherical coordinates in a channel both on the equator and in mid-latitudes we now turn in the next chapter to the application of the same idea of constructing an approximate Schrödinger equation to the wave theory on the entire spherical earth. In this problem, the boundary conditions of no flow through the channel walls do not apply and, in addition, the domain is too large for an expansion of the dependent variables near some specific latitude to yield plausible estimates.

Reference

De-Leon Y, Paldor N (2009) Linear waves in midlatitudes on the rotating spherical earth. J Phys Oceanogr 39:3204–3215. doi:10.1175/2009JPO4083.1

Chapter 6
Planetary and Inertia-Gravity Waves on the Rotating Spherical Earth

In the preceding chapters the physical setup included a channel and in these problems the boundary conditions that determine the eigensolutions (i.e., eigenvalues and eigenfunctions) of the eigenvalue problems are no-flow through the channel walls. For these boundary conditions it was natural to transform the set of two first-order equations, (2.4) for (V, η) on a plane and (4.4) for $(V \cos \phi, \eta)$ on a sphere, to a single second-order equations for V or $V \cos \phi$, respectively, i.e., in the channel setup, η was eliminated from the set of two first-order equations. In contrast, on the entire spherical earth there is no clear preference to one of the two variables and considerations other than those involving the wall boundary conditions should determine whether to eliminate η or $V \cos \phi$ in order to obtain the second-order eigenvalue equation.

The eigenvalue equations for zonally propagating waves in spherical coordinates is (4.4) for $(V \cos \phi, \eta)$ since the only assumption involved in transforming the linearized SWE set (4.1) to (4.4) is the wave structure $e^{ik(\lambda - Ct)}$ for u, V and η. Since on a sphere the boundary conditions that solutions of (4.4) have to satisfy are regularity of both $V \cos \phi$ and η at the singular poles, where $\cos \phi = 0$, one can also eliminate $V \cos \phi$ from (4.4) to obtain a single second order equation for η. Thus, in contrast to all channel problems where the Schrödinger eigenvalue problem was naturally derived by eliminating η on a sphere two eigenvalue problems can be constructed by eliminating either η or $V \cos \phi$ from system (4.4). Regardless of which of these two variables is eliminated from (4.4) the resulting second-order equation has the same generic form:

$$\alpha \frac{\partial^2 \psi}{\partial \phi^2} + \left[\omega^2 - \frac{\alpha}{C} - \left(\sin^2 \phi + \frac{\alpha k^2}{\cos^2 \phi} \right) + \alpha U_2(\phi) \right] \psi = 0, \qquad (6.1)$$

and the difference between the Schrödinger equations for $V \cos \phi$ (obtained by eliminating η) and that for η (obtained by eliminating $V \cos \phi$) is manifested only in the different expressions of $U_2(\phi)$ and the different definition of ψ. The fact that the eigenvalue equation (6.1) has the same form regardless of whether $V \cos \phi$ or η is eliminated implies that the *approximate* Schrödinger equation obtained by

© The Author(s) 2015
N. Paldor, *Shallow Water Waves on the Rotating Earth*,
SpringerBriefs in Earth System Sciences, DOI 10.1007/978-3-319-20261-7_6

neglecting $\alpha U_2(\phi)$ relative to $U_1(\phi) = \sin^2\phi + \alpha k^2/\cos^2\phi$ based on the fact that on earth $\alpha \ll 1$ is identical in the two cases and has the form:

$$\alpha\frac{\partial^2\psi}{\partial\phi^2} + \left[k^2C^2 - \frac{\alpha}{C} - \left(\sin^2\phi + \frac{\alpha k^2}{\cos^2\phi}\right)\right]\psi = 0 \tag{6.2}$$

where $k^2C^2 - \alpha/C$ is the energy, E. As before each (eigen)value of the energy, E_n, is associated with three-phase speed values that are the roots of the cubic equation $k^2C^3 - E_nC - \alpha = 0$.

Harmonic wave solutions are derived from (6.2) by eliminating the dependence on ϕ of its potential, $\sin^2\phi + \alpha k^2/\cos^2\phi$, i.e., by substituting this potential with αk^2, its constant value on the equator. The resulting equation is precisely that of Harmonic waves solved in Chap. 4 and the only difference is that the boundary conditions are applied at $\phi = \frac{1}{2}\pi$ instead of $\phi = \frac{1}{2}\delta\phi$. Thus, the eigensolutions (4.12) and (4.13) apply to the entire sphere when $\delta\phi$ is replaced by π (so $z = \phi/\pi$). Given the singularity of $1/\cos^2\phi$ at the poles these Harmonic wave solutions derived by ignoring the ϕ-dependence of the potential are of little relevance to the dynamics.

As for the more cumbersome Trapped wave solution following De-Leon and Paldor (2011), and as was shown in Chaps. 4 and 5 when η is eliminated from (4.4), ψ and U_2 are given by:

$$\psi(\phi) = \frac{V(\phi)\cos\phi}{\sqrt{\frac{\alpha}{C\cos\phi} - C\cos\phi}}$$

and

$$U_2(\phi) = \frac{3}{4}\tan^2\phi\left(\frac{\alpha + C^2\cos^2\phi}{\alpha - C^2\cos^2\phi}\right)^2 - \frac{1}{2}\left(\frac{\alpha + C^2\cos^2\phi}{\alpha - C^2\cos^2\phi}\right) - \tan^2\phi\left(\frac{\alpha + 2C^2\cos^2\phi}{\alpha - C^2\cos^2\phi}\right). \tag{6.3}$$

Similarly, when $V\cos\phi$ is eliminated from (4.4) the definition of $\psi(\phi)$ and the expression for $U_2(\phi)$ in (6.1), calculated in the same way as the expressions in (6.3), turn out to be (see Paldor et al. 2013 for more details):

$$\psi(\phi) = \eta(\phi)\sqrt{\frac{\alpha C\cos\phi}{\omega^2 - \sin^2\phi}}$$

and

$$U_2(\phi) = -\frac{1}{4}\tan^2\phi - \frac{1}{2} + \frac{3\sin^2\phi\cos^2\phi}{(\omega^2 - \sin^2\phi)^2} + \frac{1 - 3\sin^2\phi}{\omega^2 - \sin^2\phi} - \frac{2}{C}\frac{\omega^2}{\omega^2 - \sin^2\phi}. \tag{6.4}$$

Though the derivation of the approximate Schrödinger equation, (6.2), is based solely on the assumption that $\alpha \ll 1$ the validity of the resulting approximate equation has to be verified by examining the singular cases where the denominators of the neglected potentials $U_2(\phi)$ in (6.3) and (6.4) vanish. The above relations between ψ and either $V \cos \phi$ or η imply that the regularity of $V \cos \phi$ and η is guaranteed at $\phi = \pm\pi/2$ only when ψ vanishes at these points.

In addition to this polar singularity, the denominators of the neglected potentials $U_2(\phi)$ also vanish when $\omega^2 = \sin^2 \phi$ and when $\alpha = C^2\cos^2\phi$ and these two cases have to be examined. For $\omega^2 > 1$, the denominator of the three last terms of $U_2(\phi)$ in (6.4), $\omega^2 - \sin^2\phi$, never vanishes, and therefore, the neglect of $\alpha U_2(\phi)$ relative to $U_1(\phi)$ is uniformly valid in (6.4) at all latitudes in this high-frequency case. On the other hand, for $C^2 < \alpha$, the denominator of the three terms of $U_2(\phi)$ in (6.3), $\alpha - C^2\cos^2\phi$ never vanishes and therefore the neglect of $\alpha U_2(\phi)$ relative to $U_1(\phi)$ is also uniformly valid in (6.3) at all latitudes in this low-frequency case. Thus, the eigensolutions of the approximate Schrödinger equation, (6.2), are expected to provide accurate approximations for both the phase speeds (via the values of the energy, $E_n = k^2C^2 - \alpha/C$) and the meridional amplitude structure (via the eigenfunctions) of zonally propagating waves on the spherical earth for the high-frequency Inertia-Gravity waves as well as the slowly propagating Planetary waves. The above classification of the two cases in which (6.2) is uniformly valid throughout the entire sphere suggests that each of these cases corresponds to a slightly different range of α, the only non-dimensional parameter, and these ranges of α and their physical implications are described and examined separately in the remainder of this chapter.

6.1   Baroclinic (i.e., "Thin") Ocean

A sensible first step in the derivation of solution to (6.2) subject to the boundary conditions $\psi = 0$ at $\phi = \pm\pi/2$ is to simplify $U_1(\phi)$ so as to obtain a Schrödinger equation with known eigensolutions, i.e., an equation for which explicit expressions already exist for both the energy levels E_n and the corresponding eigenfunctions ψ_n. As in the channel case studied in Chap. 4, the symmetry of $U_1(\phi)$, i.e., $U_1(-\phi) = U_1(\phi)$ implies that its power series (e.g., Taylor series) expansion contains only even powers of ϕ. Retaining only second-order terms in the expansions of $\sin^2 \phi$ and $1/\cos^2 \phi$ transforms Eq. (6.2) into the known Schrödinger equation of Harmonic Oscillator of Quantum Mechanics that is identical to the differential equation (but with different boundary conditions) derived in (4.10):

$$\alpha\frac{\partial^2 \psi}{\partial \phi^2} + \left[E - \alpha k^2 - \left(1 + \alpha k^2\right)\phi^2\right]\psi = 0. \tag{6.5}$$

The eigensolutions derived in Chap. 4 to this equation are also relevant to the whole sphere but the range of validity of (6.5) and the effect of the boundary conditions of regularity at the poles instead of the no normal flow conditions at the channel walls need to be examined. The eigensolutions of (6.5) include the energy levels, E_n, and eigenfunctions, ψ_n, given by:

$$E_n = \alpha k^2 + (2n+1)\sqrt{\alpha + \alpha^2 k^2}$$

and   (6.6)

$$\psi_n = H_n\left(\phi/\sqrt{\gamma}\right)e^{-\phi^2/(2\gamma)}, \quad n = 0, 1, 2, \ldots$$

where H_n are Hermite polynomials of order n and $\gamma = \left(\frac{\alpha}{1+\alpha k^2}\right)^{1/2}$.

Each of the energy levels in (6.6) is associated with a special latitude known as the turning latitude, denoted here by ϕ_{turn}, where the coefficient of ψ in (6.5), $E_n - \alpha k^2 - (1 + \alpha k^2)\phi_{\text{turn}}^2$, changes sign from positive at $\phi < \phi_{\text{turn}}$ to negative at $\phi > \phi_{\text{turn}}$. Substituting the expression for E_n, (6.6), into the coefficient of ψ in (6.5) and equating this coefficient to zero yield the following explicit expression for ϕ_{turn}:

$$(\phi_{\text{turn}})^2 = (2n+1)\frac{\alpha^{1/2}}{(1+\alpha k^2)^{1/2}}.$$ (6.7)

In accordance with the change of sign of the coefficient of ψ in (6.5) at $\phi = \phi_{\text{turn}}$, the eigenfunction corresponding to E_n, i.e., $\psi_n(\phi)$, changes its behavior there from oscillatory (but not necessarily periodic) at $\phi < \phi_{\text{turn}}$ to decaying/increasing (possibly faster than exponential) at $\phi > \phi_{\text{turn}}$.

Solutions of the differential Eq. (6.5) are eigenfunctions of the eigenvalue problem only when they satisfy the boundary condition $\psi_n(\phi = \pm\pi/2) = 0$, and therefore only the solutions that decay to zero in the ranges $\phi_{\text{turn}} < \phi < \pi/2$ and $-\phi_{\text{turn}} > \phi > -\pi/2$ are acceptable as approximate eigenfunctions of (6.2). Since the rate at which these solutions decay to zero with latitude at $|\phi| > |\phi_{\text{turn}}|$ is fast, given by the exponent of $-\phi^2/(2\gamma)$ which is close to $-\phi^2/(2\alpha^{1/2})$ for $\alpha k^2 \ll 1$, the maximal latitude where these solutions are nonzero is very well approximated by ϕ_{turn}. Thus, the parabolic expansion of $U_1(\phi)$ that enabled the transformation of (6.2)–(6.5) is only valid when $\phi_{\text{turn}} < 1$, so according to (6.7) the validity of (6.5) is limited by:

$$\phi_{\text{turn}} = \frac{2n+1}{\sqrt{\frac{1}{\alpha}+k^2}} < 1, \quad n = 0, 1, 2, \ldots$$ (6.8)

This condition is more restrictive than the $a < 1$ condition employed to justify the neglect of $\alpha U_2(\phi)$ from (6.1) since for any values of α and k it sets an upper bound on the mode number, n, below which the approximate solution (6.6) is valid. In a baroclinic ocean $\alpha = 5 \times 10^{-6}$ (this is also the value of α for a 0.5 m deep layer of fluid) and the foregoing theory applies to $n < O(100)$. In the atmosphere where the speed of gravity waves is an order of magnitude larger than in a baroclinic ocean, the value of α is 10^{-4} so the corresponding bound on n is $n < O(10)$. In a barotropic ocean where $(gH)^{1/2}$ is about 200 m/s the value of α is 0.05 and the bound on n is 1 since at this value of α (6.8) is violated even for $n \geq 2$ so the foregoing theory is expected to apply only to $n = 0$ and $n = 1$. Large enough values of k can alter these bounds somewhat, particularly at small α^{-1}.

The analytical results derived in this section serve as a guide in the interpretation of numerical solutions of the original exact differential system (4.2) for $-\pi/2 \leq \phi \leq \pi/2$ subject to the boundary conditions of regularity of u, V and η at $\phi = \pm\pi/2$. The numerical solutions described below are also intended to validate the analytical results and provide exact quantitative (and not just order of magnitude) estimates for the parameter values where the results of analytical theory of a "thin" ocean apply.

Figure 6.1, adapted from De-Leon and Paldor (2011), verifies the approximations described above that involve both the neglect of $\alpha U_2(\phi)$ and the limitation on n that results from the parabolic approximation of $U_1(\phi)$. Three values of α are used: 10^{-2}, 10^{-4} and 10^{-6} and in all three cases, the zonal wavenumber was set to $k = 10$. The three potentials, $U_1(\phi) + \alpha U_2(\phi)$, $U_1(\phi)$ and the parabolic approximation to $U_1(\phi)$—$\alpha k^2 + (1 + \alpha k^2)\phi^2$, shown in the left panels, clearly demonstrate that in all three cases $U_1(\phi)$ provides an accurate approximation to $U_1(\phi) + \alpha U_2(\phi)$ throughout most of the $-\pi/2 < \phi < \pi/2$. The two potentials differ from one another only in the immediate vicinity of the poles at latitudes far beyond ϕ_{turn}—the latitude where the eigenfunction $\psi_n(\phi)$ in (6.6) begins its faster-than-exponential decay and where the parabolic potential $\alpha k^2 + (1 + \alpha k^2)\phi^2$ ceases to provide a valid approximation to $U_1(\phi)$.

Thus, even for $\alpha = 10^{-2}$ (top row), the neglect of $\alpha U_2(\phi)$ does not invalidate the foregoing theory. On the other hand, the approximation of $U_1(\phi)$ by $\alpha k^2 + (1 + \alpha k^2)\phi^2$ is limited to low latitudinal mode numbers, n, that satisfy (6.8) and for $\alpha = 10^{-2}$ the theory applies to $n < 2$ only since for $n \geq 2$ ϕ_{turn} on the RHS of (6.8) is not sufficiently smaller than 1. The panels on the right column of Fig. 6.1 show a zoom in of the three potentials on the $\phi < 0.6$ range where the approximation of $U_1(\phi)$ by $\alpha k^2 + (1 + \alpha k^2)\phi^2$ should hold. Also shown are a few representative energy levels associated with sufficiently small ϕ_{turn}. A conservative estimate of the number of energy levels contained in this ϕ-range is provided by the index of the highest energy level shown in each panel and this estimate can be somewhat increased for a less conservative, but still fairly accurate, estimate.

The dispersion relations of Planetary waves and westward propagating Inertia-Gravity waves derived from these estimates of the energy levels by combining them with the definition of the energy in terms of the waves' phase speeds, $E = k^2 C^2 - \alpha/C$, are shown in Fig. 6.2 for two values of α and for mode numbers $n = 1, 3, 7$ (Planetary waves) and $n = 1$–8 (Inertia-Gravity). These calculations demonstrate the high accuracy of the analytic estimates for sufficiently low values of αn^2 and a slight error can be detected in the $n = 8$ curve of $\alpha = 10^{-4}$ panel where αn^2 equals $0.0064 = 0.64$ %.

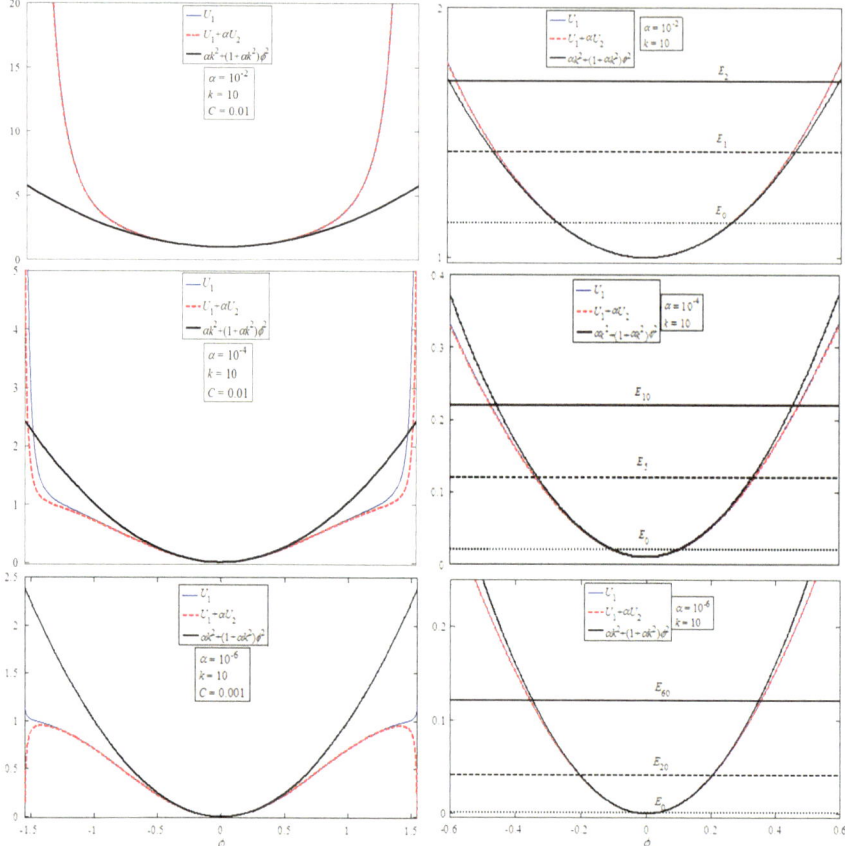

Fig. 6.1 *Left column* The exact and approximate potentials. The approximations of $U_1(\phi) +$ $\alpha U_2(\phi)$ (*red dashed line*) by $U_1(\phi)$ (*thin solid blue line*) are not accurate near the poles. The approximations of $U_1(\phi)$ by $\alpha k^2 + (1 + \alpha k^2)\phi^2$ (*thick solid line*) are not accurate at latitudes larger than about 0.6. *Right column* Zoom in on the potentials in the range $-0.6 \leq \phi \leq 0.6$ and several the energy levels for which $\phi_{turn} < 0.6$. As α decreases going from top to bottom the number of energy levels with turning latitudes $\phi_{turn} < 0.6$ increases: For $\alpha = 10^{-2}$ $\phi_{turn} < 0.6$ only for $n \leq 2$ (*upper panel*); For $\alpha = 10^{-4}$ $\phi_{turn} < 0.6$ for $n < 10$ (*middle panel*) while for $\alpha = 10^{-6}$ $\phi_{turn} < 0.6$ even for $n = 60$ (*lower panel*). Permission from Co-Action/TellusA: doi:10. 1111/j.1600-0870.2010.00490.x

6.2   Barotropic (i.e., "Deep") Ocean

As discussed above, in a barotropic ocean where H is about 4 km the value of α is about 0.05 which is sufficiently small compared to 1 to justify the neglect of $\alpha U_2(\phi)$ relative to $U_1(\phi)$ so that the assumption needed to transform (6.1) to (6.2) is valid at this value of α. Indeed, as expected, the numerically calculated graphs of $U_1(\phi)$ and those of $U_1(\phi) + \alpha U_2(\phi)$ shown in Fig. 6.1 for $\alpha = 0.01$ are very close

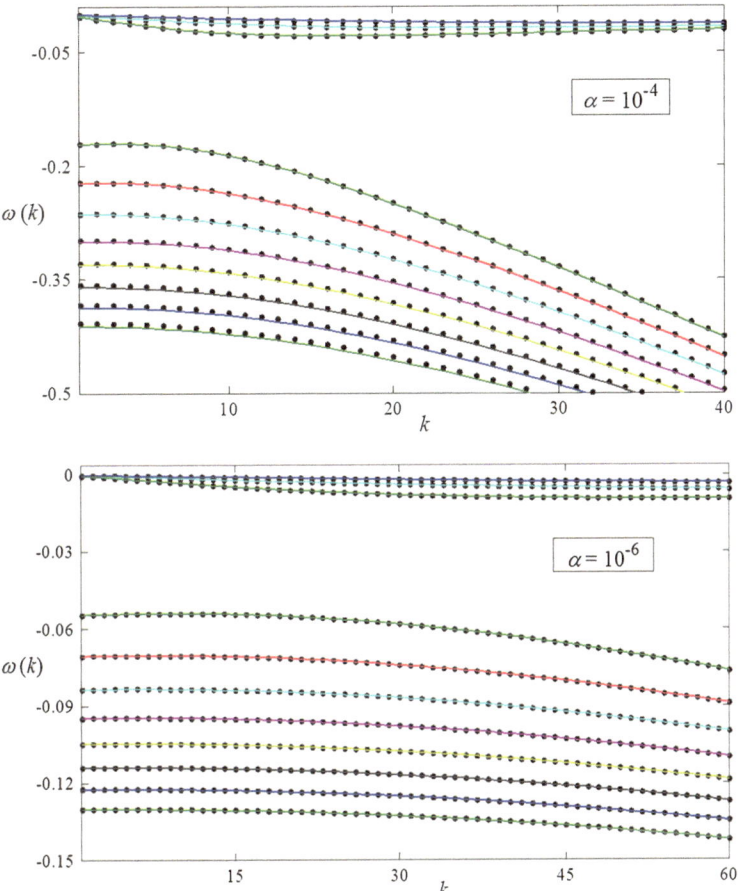

Fig. 6.2 The dispersion relations of mode numbers 1, 3, and 7 of Planetary waves and mode numbers 1–8 of westward propagating Inertia-Gravity waves for the indicated values of α. Permission from Co-Action/TellusA: doi:10.1111/j.1600-0870.2010.00490.x

to one another except for the immediate vicinity of the poles. However, according to the constraint (6.8) at this value of α (6.6) provides acceptable approximate eigensolutions of (6.2) only for $n = 0$ and 1 while at higher meridional mode numbers the solutions of this equation have to be derived by solving it (including the associated boundary conditions) directly since no exact or approximate solutions have been previously derived. The derivation of these sought solutions of the Schrödinger Eq. (6.2) when αn^2 is not small follows the method described in Paldor et al. (2013).

   The starting point of the derivation of new solutions to (6.2) is handling the singularity of $U_1(\phi)$ at the poles. The development of the theory is facilitated by defining $\mu = \sin \phi$ as the new independent variable of (6.2) in terms of which the equation transforms to:

$$\alpha\left(1 - \mu^2\right)\frac{\partial^2 \psi}{\partial \mu^2} - \alpha\mu\frac{\partial \psi}{\partial \mu} + \left[k^2 C^2 - \frac{\alpha}{C} - \left(\mu^2 + \frac{\alpha k^2}{1 - \mu^2}\right)\right]\psi = 0. \tag{6.9}$$

The polar singularity of (6.2) appears in this equation as the $\mu = \pm 1$ singularity. In order to address this singularity the sought eigenfunction is decomposed into $\psi(\mu) = (1 - \mu^2)^\delta G(\mu)$ where the power δ has yet to be determined and where $G(\mu)$ is assumed to be a regular function that has a Taylor series expansion. The function $(1 - \mu^2)^\delta$ that multiplies the regular function $G(\mu)$ is the counterpart of the Gaussian that multiplies Hermite polynomials in the Hermite function solutions of the previous subsection. These two "envelope" functions guarantee that the eigensolutions satisfy the regularity boundary conditions at the poles so $G(\mu)$ (and in the case of a baroclinic ocean the Hermite polynomials) has to satisfy only the differential equation, thus determining the energy levels.

Substituting the assumed form $\psi(\mu) = (1 - \mu^2)^\delta G(\mu)$ in (6.9) and dividing the resulting equation by $\alpha(1 - \mu^2)^{\delta-1}$ yields the differential equation:

$$\left(1 - \mu^2\right)^2 G'' - 4\delta\mu\left(1 - \mu^2\right)G' + 4\delta(\delta - 1)\mu^2 G - 2\delta\left(1 - \mu^2\right)G$$
$$+ 2\delta\mu^2 G - \mu\left(1 - \mu^2\right)G' + \frac{E - \mu^2}{\alpha}\left(1 - \mu^2\right)G - k^2 G = 0, \tag{6.10}$$

where $E = k^2 C^2 - \alpha/C$. For regular G, G' and G'' setting $\mu^2 = 1$ in this equation yields the quadratic equation for δ:

$$\left(4\delta(\delta - 1) + 2\delta - k^2\right)G = 0. \tag{6.11}$$

The positive root of this equation, which is associated with a regular $\psi(\mu)$ at $\mu^2 = 1$, is:

$$2\delta = \frac{1}{2} + \sqrt{\frac{1}{4} + k^2}. \tag{6.12}$$

Since $\sqrt{\frac{1}{4} + k^2} > \max\{\frac{1}{2}, k\}$ Eq. (6.12) guarantees that $2\delta \geq \max\{1, k\}$ for all k and that this inequality is strict for $k > 0$. Since $(1 - \mu^2)^\delta = \cos^{2\delta} \phi$ it turns out that $\psi(\phi)$ decays to 0 at the poles faster than $\cos^k \phi$ for all $k \geq 1$ and as $\cos\phi$ for $k = 0$.

Having determined the "envelope" function $(1 - \mu^2)^\delta$ we now turn our attention to $G(\mu)$. Using (6.11) to substitute $(4\delta(\delta - 1) + 2\delta)G$ for $k^2 G$ (the last term on the LHS) of (6.10) and dividing the resulting equation by $(1 - \mu^2)$ yields the following regular equation for $G(\mu)$:

$$G'' - \mu^2 G'' - (1 + 4\delta)\mu G' + \left(\frac{E}{\alpha} - 4\delta^2\right)G - \frac{1}{\alpha}\mu^2 G = 0. \tag{6.13}$$

Although this equation is regular, in its present form it does not have polynomial solutions of finite degree N. To see why this is the case, assume that μ^N is the highest power in the solution, in which case the $\mu^2 G$ term (the last term on its LHS) is proportional to μ^{N+2} while all other terms have powers of μ smaller than, or equal to, N. Thus, this μ^{N+2} term cannot be balanced by any other term in the equation so it has to be equal to zero which contradicts the initial assumption that the solution is a polynomial of degree N. Since $\alpha < 1$, the coefficient of the $\mu^2 G$ term in (6.13), $1/\alpha$, is larger than 1 so it is unclear how to derive explicit expressions for the solutions of this equation.

An alternative way of finding explicit expressions for the solutions of (6.13) is to expand $G(\mu)$ as an infinite Taylor series. We thus let $G(\mu) = \sum_{j=0}^{\infty} b_j \mu^j$ where the coefficients $\{b_j, j = 0, 1, 2, \ldots\}$ are to be determined by substituting this expansion into (6.13). Collecting like powers of μ from all terms of this equation yields the infinite series:

$$
2b_2 + \left(\frac{E}{\alpha} - 4\delta^2 \right) b_0 + \left[6b_3 + \left(-(1+4\delta) + \frac{E}{\alpha} - 4\delta^2 \right) b_1 \right] \mu
$$
$$
+ \sum_{j=2}^{\infty} \left[(j+2)(j+1)b_{j+2} + \left(-j(j-1) - j(1+4\delta) + \frac{E}{\alpha} - 4\delta^2 \right) b_j - \frac{1}{\alpha} b_{j-2} \right] \mu^j = 0.
$$

$$(6.14)$$

Equating to zero the coefficients of μ^j, $j = 0, 1, 2 \ldots$ in this equation yields the recursion relations:

$$
b_2 = \frac{4\delta^2 - \frac{E}{\alpha}}{2} b_0,
$$
$$
b_3 = \frac{(1+2\delta)^2 - \frac{E}{\alpha}}{6} b_1,
$$

$$(6.15)$$

$$
\cdot
$$
$$
\cdot
$$

$$
b_{j+2} = \frac{\left[(j+2\delta)^2 - \frac{E}{\alpha} \right] b_j + \frac{1}{\alpha} b_{j-2}}{(j+2)(j+1)}, \quad j \geq 2.
$$

The energy levels are determined by requiring the coefficient of b_n on the RHS of the last general relation of this equation to vanish at some large n, i.e., $E_n = \alpha(n + 2\delta)^2$. For the infinite series $G(\mu) = \sum_{j=0}^{\infty} b_j \mu^j$ to converge at $\mu = \pm 1$, the ratio b_{j+2}/b_j has to be smaller than 1 and independent of j in the $j \to \infty$ limit. Denoting $X = b_{n+2}/b_n \approx b_n/b_{n-2}$ the last relation in (6.15) implies $X = \frac{1}{n\sqrt{\alpha}}$ so $X \xrightarrow{n \to \infty} 0$ regardless of the value of α and the series converges rapidly for $j > n$. The accuracy of this estimate of the energy levels:

$$E_n = \alpha(n + 2\delta)^2, \quad n \geq 0, \tag{6.16}$$

has to be determined numerically for various combinations of α and corresponding n values since the derivation of this estimate is based on the assumption that $\alpha n^2 \gg 1$.

Since the eigenvalue equation, (6.9), is invariant under the change-of-sign of the independent variable, μ, the solutions must be either symmetric or anti-symmetric. The eigenfunction corresponding to E_n is proportional to $G_n = \sum_{j=0}^{n} b_j \mu^j$ where the b_j's are given by the recursion relation (6.15) with E_n substituted for E. The symmetry of G_n is determined by the value of n, so $G_n(\mu)$ is an even/odd function of μ when n is even/odd. The even solutions are obtained by letting $b_0 = 1$ (the value of 1 is an arbitrary amplitude) and $b_1 = 0$ in (6.15) and the odd solutions by letting $b_0 = 0$ and $b_1 = 1$. From their definitions it is obvious that $\psi(\mu), V(\mu)$ and $v(\mu)$ in the case of (6.3) and $\psi(\mu)$ and $\eta(\mu)$ in the case of (6.4) have the same symmetry as G_n.

In the remainder of this section the theoretical results obtained above will be applied to derive explicit expressions for the dispersion relations of barotropic Planetary and Inertia-Gravity waves and for the meridionally dependent amplitudes associated with these relations.

6.2.1 Dispersion Relation of Planetary Waves

The phase speed of Planetary waves, approximated by $C_n = -\alpha/E_n$, can be combined with (6.16) to yield the dispersion relation of these waves in the $n^2 \to \infty$ limit:

$$C_n = -\frac{\alpha}{E_n} = \frac{-1}{(n + 2\delta)^2} \tag{6.17}$$

where $2\delta = 1/2 + (1/4 + k^2)^{1/2}$ according to (6.12) i.e., δ is a function of k only so C_n is independent of α.

The results shown in Fig. 6.3 (the figures in this subsection were reproduced from Paldor et al. (2013)) clearly demonstrate the accuracy of the analytical estimates of E_n that were formally derived in (6.16) only for $n \to \infty$ but seem to be accurate even for $n = 5$, i.e., when $\alpha n^2 = 1.25$, which is only marginally larger than 1. It should not come as a surprise that for small n (and k) values, where the condition $n^2 \to \infty$ is violated, the error of the analytic result is as high as 50 %. The frequency of all modes was set to 0 at $k = 0$ (i.e., no calculations were performed for $k = 0$).

As is evident from (1.7) in planar Planetary wave theory the phase speed C is independent of α (identified by gH) at wavelengths much shorter than the radius of deformation (proportional to $\alpha^{1/2}$ in the scaling used here). The results derived in (6.17) regarding the independence of C_n on α when $n \to \infty$ (and as long as $\alpha < 1$ to

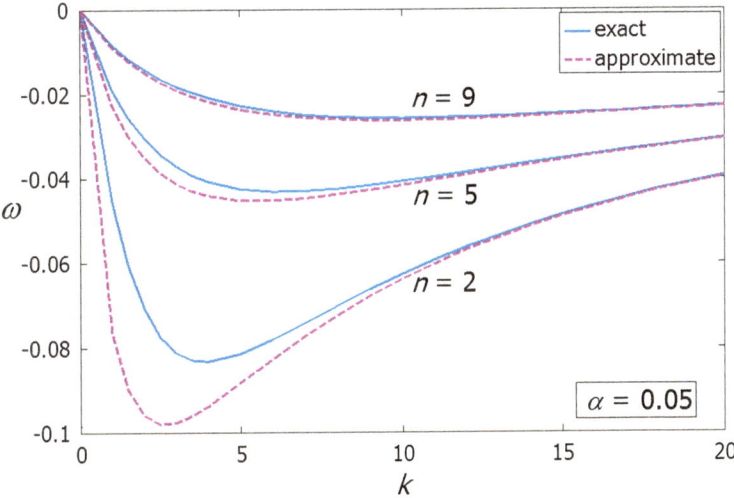

Fig. 6.3 The dispersion relation, $\omega = kC$ as a function of k, of Planetary waves. Approximate expression from (6.17) (*purple dashed curves*) and the exact C calculated numerically from (4.2) (*blue solid curves*). Permission from Cambridge University Press: J. Fluid Mech. doi:10.1017/jfm. 2013.219

ensure that αU_2 can be neglected) extend this planar result to spherical geometry but with a modified combination of n and k. This conclusion is confirmed in Fig. 6.4 where the exact C-values calculated numerically from (4.2) are compared to the values given in (6.17) for $n = 15 = k$. As the value of $\alpha(n + 2\delta(k))^2$ increases from

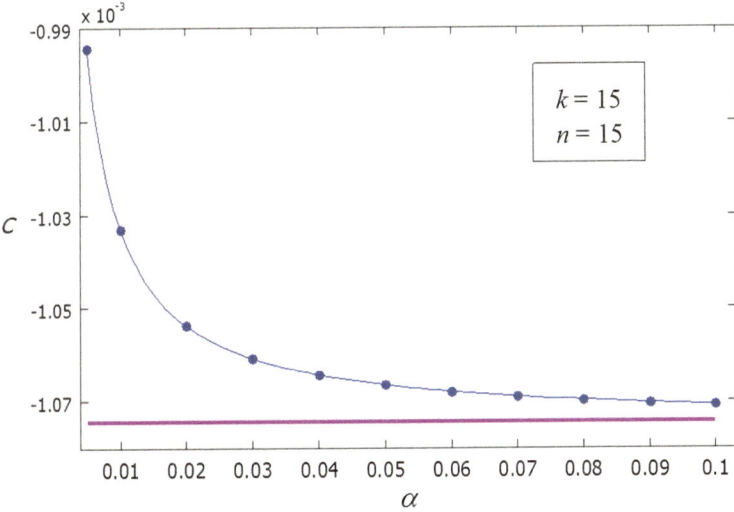

Fig. 6.4 The asymptotic approach of the exact value of C (*blue dotted curve*) to the value given by (6.17) (*solid constant line* at C = −1.0744) as $\alpha(n + 2\delta(k))$ is increased. Permission from Cambridge University Press: J. Fluid Mech. doi:10.1017/jfm.2013.219

2.25 at $\alpha = 0.005$ to 90 at $\alpha = 0.1$, the corresponding relative deviation of C from the asymptotic value (6.17) decreases from about 0.08 to 0.003. The validity of the $n^2 \to \infty$ theory is bolstered by the small (8 %) deviation of the value of C at $\alpha(n + 2\delta(k))^2 = 2.25$ as well as by its monotonic approach to the value given by (6.17) when $\alpha(n + 2\delta(k))^2$ is increased.

6.2.2 Dispersion Relation of Inertia-Gravity Waves

The phase speed of Poincaré waves, approximated by $C_n = \pm(E_n)^{1/2}/k$, can be combined with the expression for the eigenvalue, E_n, found in (6.16) to yield the following dispersion relation of Poincaré waves which is valid in the limit when $n^2 \to \infty$:

$$C_n = \pm \frac{\sqrt{E_n}}{k} = \pm \frac{\sqrt{\alpha}(n + 2\delta)}{k}. \qquad (6.18)$$

The validation of this dispersion relation of the positive Inertia-Gravity waves is shown in Fig. 6.5 (the results for negative Inertia-Gravity waves are similar) where this expression is compared with exact numerical calculations of $C(k, n; \alpha = 0.05)$. As expected, the accuracy of the analytic approximate expression for C_n given in (6.18) increases with $(n + 2\delta(k))$. The analytic solution (6.18) of the eigenvalue equation, (6.9), is valid only when the following two conditions are satisfied: (1) The term $\psi/\cos^2(\phi) = \cos^{2(\delta-1)}(\phi)G(\sin(\phi))$ is regular, i.e., only for $k > 1$, and 2) the frequency $\omega = Ck$ is larger than 1, so the neglect of αU_2 in (6.4) is valid in $-\pi/2 \leq \phi \leq \pi/2$. However, although the associated eigenfunctions obtained

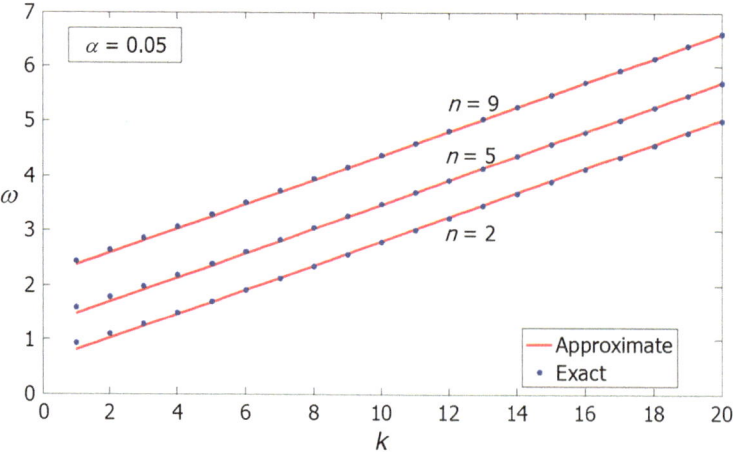

Fig. 6.5 The dispersion relations, $\omega = Ck$ as a function of k, of the positive Inertia-Gravity wave. Exact values of C (*points*) calculated numerically from (4.2); approximate values of C (*solid lines*) from (6.18). Permission from Cambridge University Press: J. Fluid Mech. doi:10.1017/jfm.2013.219

numerically for $k = 1$ are singular at high latitudes, i.e., condition 1 is violated (but $\eta(\phi)$ calculated from system (4.2) is regular), (6.18) yields fairly accurate approximations to C. This fit of the eigenvalues (and phase speeds) is accurate even for the $n = 2, k = 1$ mode where the frequency is smaller than 1, i.e., condition 2 is violated.

6.2.3   Eigenfunctions

In addition to the approximate energy levels that were compared to their exact counterparts in the preceding subsections, the analytic approximate eigenfunction $\psi(\mu) = (1 - \mu^2)^\delta G(\mu)$, where $G(\mu)$ is the polynomial approximation, truncated at the chosen n and satisfying the recursion relation (6.15) with the b_{j-2} term omitted from all expressions of b_{j+2} for $2 < j < n - 2$ should also be compared to their exact counterparts. These exact solutions of the eigenfunctions can be derived from numerical solutions of (4.2) that for Planetary waves can be employed to yield ψ from the V field by applying the $V(\psi)$ transformation (6.3), while for Inertia-Gravity waves, the exact numerical solution of (4.2) for η yields ψ by the $\eta(\psi)$ transformation (6.3). The comparisons shown in Fig. 6.6 clearly demonstrate that the accuracy of the analytic results is not limited to the eigenvalues only but prevails also in the eigenfunctions. At (k, n) values higher than $(5, 5)$, the exact and approximate functions cannot be distinguished from one another.

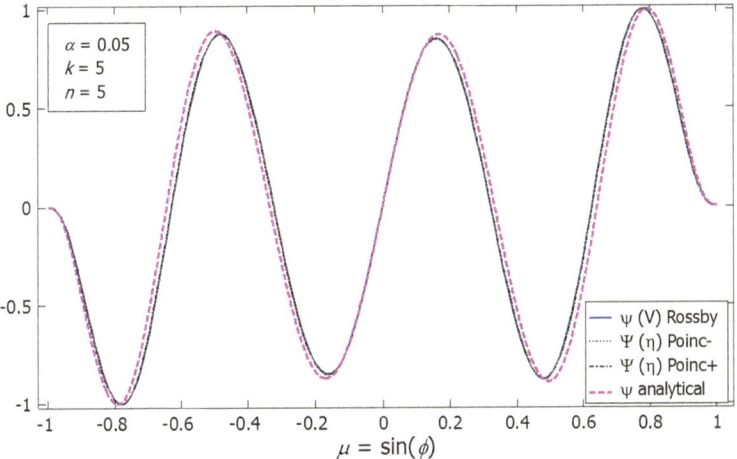

Fig. 6.6 The exact (*solid, dotted, dashed-dotted lines*) eigenfunctions $\psi(\mu)$ of the three wave types and the single approximate analytic $\psi(\mu)$ (*dashed line*). All eigenfunctions are normalized so the maximum value is 1. The three exact functions of the indicated n and k cannot be distinguished from one another while the approximate analytic function deviates very slightly from them. Permission from Cambridge University Press: J. Fluid Mech. doi:10.1017/jfm.2013.219

References

De-Leon Y, Paldor N (2011) Zonally propagating wave solutions of laplace tidal equations in a baroclinic ocean of an aqua-planet. Tellus 63A:348–353. doi:10.1111/j.1600-0870.2010.00490.x

Paldor N, Shamir O, De-Leon Y (2013) Planetary (Rossby) waves and inertia-gravity (Poincaré) waves in a barotropic ocean over a sphere. J Fluid Mech 726:123–136

Chapter 7
Kelvin Waves on the Rotating Spherical Earth

As was noted in Chaps. 4–6, where Planetary and Inertia-Gravity waves were derived on a sphere, Kelvin waves cannot be derived in spherical coordinates by simply setting $C^2 = \alpha$ since even with this substitution the coefficient of η in the equation for $\partial(V\cos\phi)/\partial\phi$ [the upper right term in the matrix on the RHS of (4.4)] does not vanish, so the two first-order equations remain coupled. This is in contrast to Cartesian coordinates (both in mid-latitudes and on the equator) where the substitution $C^2 = \alpha$ decouples the two equations by setting the coefficient of η in the equation for $\partial V/\partial y$ [the second term in the top line on the RHS of (2.4)] equal to zero.

This straightforward analytical assertion is supported by numerical integration of system (4.4), which is exact for zonally propagating wave solutions, with $C = \sqrt{\alpha}$ over the sphere. Letting $C = \sqrt{\alpha}$ and $\mu = \sin\phi$ and defining $v(\mu) \equiv V\cos\phi$ transforms system (4.4) to:

$$\alpha^{1/2}\left(1 - \mu^2\right)\frac{\partial}{\partial\mu}\begin{pmatrix}\eta \\ v\end{pmatrix} = \begin{pmatrix}-\mu & k^2 - \frac{\mu^2}{\alpha} \\ \alpha\mu^2 & \mu\end{pmatrix}\begin{pmatrix}\eta \\ v\end{pmatrix}. \tag{7.1}$$

Upon expanding the solutions of these equations near the singular poles $\mu = \pm 1$ it becomes evident that the regular solutions at the poles are given by:

$$\eta \xrightarrow{\mu^2 \to 1}\left(1 - \mu^2\right)^{k/2}; \quad v \xrightarrow{\mu^2 \to 1}\frac{-\mu\alpha}{1 + k\sqrt{\alpha}}\left(1 - \mu^2\right)^{k/2}. \tag{7.2}$$

If $C = \sqrt{\alpha}$ is indeed an "eigenvalue" of system (4.4) then system (7.1) should have solutions that are differentiable throughout $-1 < \mu < +1$ and regular at $\mu = \pm 1$. However, when system (7.1) is integrated numerically twice: the first time from near $\mu = 1$ to $\mu = 0$ and the second time from near $\mu = -1$ to $\mu = 0$ (i.e., away from the singular points) and the two calculated values of $\eta(\mu)$ at $\mu = 0$ are matched to yield $\eta(\mu)$ that is continuous throughout it turns out that $\partial\eta/\partial\mu$ and hence $V(\phi)\cos\phi$ are discontinuous at the matching point $\mu = 0$. An example of these calculations is shown in Fig. 7.1 and the same calculations were repeated for a large number of (α, k) values and in all of them the results were identical—no continuous solutions of system (7.1) that satisfy the regularity condition (7.2) were found.

© The Author(s) 2015
N. Paldor, *Shallow Water Waves on the Rotating Earth*,
SpringerBriefs in Earth System Sciences, DOI 10.1007/978-3-319-20261-7_7

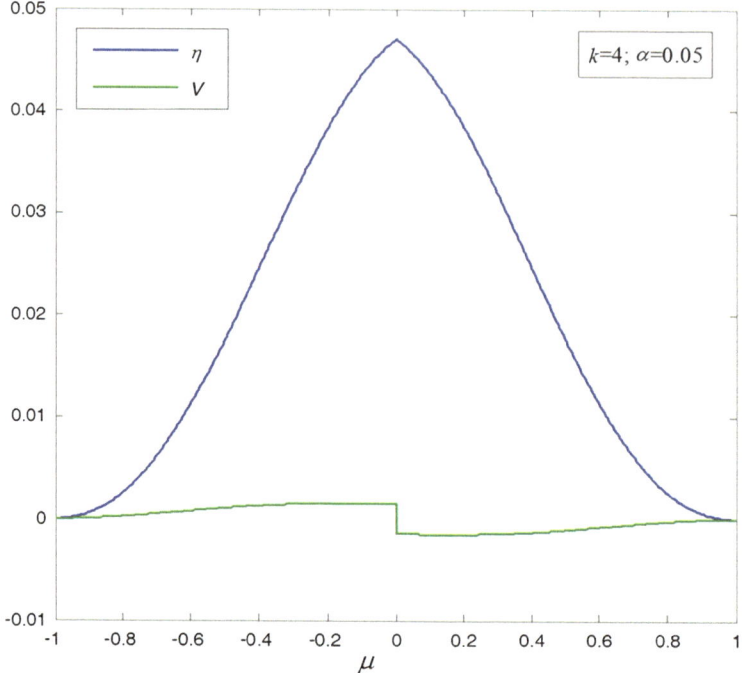

Fig. 7.1 A numerical integration of system (7.1) starting with the regular solutions (7.2) at $\mu = -1$ in the first integration and at $\mu = 1$ in the second integration. The two integrations are matched at $\mu = 0$ to ensure that $\eta(\mu)$ is continuous. As is evident from these results, the integrations lead to discontinuous $\partial\eta/\partial\mu$ (and $v(\mu) = V(\phi)\cos\phi$)

In contrast to these results on a sphere the application of this numerical method to a channel on the equatorial β-plane where $V = 0$ at the channel walls, namely integrating the governing equations (i.e., (2.4) with $C = \sqrt{\alpha}$ and $\phi_0 = 0$) from the two walls to the equator and matching the two values of η there always, yields the analytical continuous eigenfunctions where η is a Gaussian and $V = 0$ everywhere. The successful application of this numerical methodology on the equatorial β-plane validates the conclusion derived from its application to (7.1) subject to the boundary conditions (7.2) that no waves with $C = \sqrt{\alpha}$ exist on a sphere.

Having demonstrated both analytically and numerically that unlike Cartesian coordinates there is no solution to the SWE on a sphere with either $V = 0$ or $C = \sqrt{\alpha}$, i.e., there are no Kelvin waves on a sphere, we still have to consider the possibility that "Kelvin" waves exist on sphere but only as an asymptotic limit in some small parameters such as α or C. In such an asymptotic theory both $V\cos\phi$ (or v) and $C^2 - \alpha$ are negligible only at the leading order. Indeed, such an asymptotic theory for linear "Kelvin" waves on a sphere was developed by Boyd and Zhou (2008) based on a simplified version of (4.4) obtained by employing a smart "spherical equatorial" dynamics. However, since the approximate equation solved by Boyd and Zhou (2008) is also a genuine second-order eigenvalue equation it is unclear whether the solution derived is a "Kelvin" wave or the $n = 0$ Inertia-Gravity mode. As we shall shortly see, the $n = 0$ mode of the Inertia-Gravity waves is non-dispersive and passes through the origin of the $\omega(k)$ relation (but with phase speed, i.e., the slope of the $\omega(k)$ curve, larger than $\sqrt{\alpha}$), so care should be exercised in the development of such a "Kelvin" wave theory to ensure that the solution obtained does not reproduce the $n = 0$ Inertia-Gravity mode.

The subtlety of the issue at hand is demonstrated by examining the dispersion relations of the first 8 modes with positive frequencies obtained from numerical solutions of system (4.2). The dispersion relations are shown in Fig. 7.2 and the curve of the mode with the lowest frequency highlights the dilemma regarding "Kelvin" waves on a rotating sphere: Is this lowest curve a "Kelvin" wave or the $n = 0$ mode of the eastward propagating Inertia-Gravity waves? Obviously, if the answer to this question is the latter of these two possibilities then no "Kelvin" waves exist on a sphere since the frequency of Kelvin waves is lower than that of the $n = 0$ Inertia-Gravity mode.

To resolve this issue we consider the exact second-order eigenvalue Eq. (6.1) derived in Chap. 6 for any zonally propagating SW wave on a sphere by eliminating either $V\cos\phi$ or η from the two first-order equations—(4.4). Since to leading order V has to vanish in "Kelvin" waves while η does not, the leading order dynamics of Kelvin waves are describable by the η-equation, i.e., the equation obtained when $V\cos\phi$ is eliminated from (4.4) (see also Eq. (6.4)):

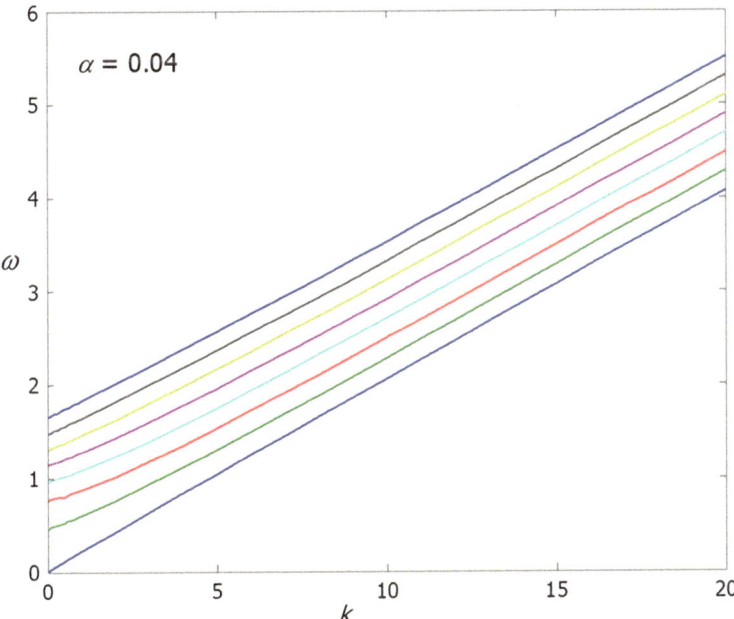

Fig. 7.2 The dispersion curves of the first eight, positive frequency, modes of system (4.2)

$$0 = \alpha \frac{\partial^2 \psi}{\partial \phi^2} + \left[\omega^2 - \frac{\alpha}{C} - \left(\sin^2 \phi + \frac{\alpha k^2}{\cos^2 \phi} \right) \right.$$

$$+ \alpha \left(-\frac{1}{4} \tan^2 \phi - \frac{1}{2} + \frac{3 \sin^2 \phi \cos^2 \phi}{\left(\omega^2 - \sin^2 \phi \right)^2} \right.$$ (7.3)

$$\left. \left. + \frac{1 - 3 \sin^2 \phi}{\omega^2 - \sin^2 \phi} - \frac{2}{C} \frac{\omega^2}{\omega^2 - \sin^2 \phi} \right) \right] \psi,$$

where ψ is related to η by:

$$\psi(\phi) = \eta(\phi) \sqrt{\frac{\alpha C \cos \phi}{\omega^2 - \sin^2 \phi}}.$$

If "Kelvin" waves exist on a sphere, then they ought to be special solutions of Eq. (7.3) that do not belong to the numerable series of regular solutions derived in Chap. 6. Such special (or degenerate) solutions can only be associated with singular cases in which the transformation leading to (7.3) is singular, i.e., only for $\omega = kC < 1$, which for $C \approx \sqrt{\alpha}$ is only relevant to the limited zonal wavenumber range $k < 1/\sqrt{\alpha}$. This finite wavelength cutoff for the possible existence of special solutions to (7.3) rules out the existence of a non-dispersive "Kelvin" wave on a sphere and ensures that the continuous non-dispersive lowest dispersion curve in Fig. 7.2 is the $n = 0$ mode of the eastward propagating Inertia-Gravity waves.

Additional independent arguments leading to the same conclusion are obtained by considering analytic approximations to the numerically calculated dispersion curves in Fig. 7.2 and by examining the eigenfunctions associated with these curves. A starting point for the analytical estimates is a derivation of the exact dispersion relations for gravity waves on a non-rotating sphere. Without rotation, the dimensional vectorial (i.e., geometry free) LSWE takes the form:

$$\frac{\partial \underline{V}}{\partial t} = -g \underline{\nabla} \eta,$$

$$\frac{\partial \eta}{\partial t} = -H \underline{\nabla} \cdot \underline{V}.$$ (7.4)

Differentiating the second equation WRT t and substituting the first equation yields:

$$\frac{\partial^2 \eta}{\partial t^2} = -H \underline{\nabla} \cdot \frac{\partial \underline{V}}{\partial t} = g H \underline{\nabla} \cdot \underline{\nabla} \eta = g H \Delta \eta$$ (7.5)

where $\Delta = \underline{\nabla} \cdot \underline{\nabla}$ (i.e., the DIV operator applied to the GRAD of a scalar function) is the Laplacian. On a sphere of radius a, the eigenfunctions of the Laplacian are the Spherical Harmonics and the eigenvalues are $-l(l+1)/a^2$, where $l = n + k \geq 0$ is the total wavenumber. Thus, the frequencies of (gravity) waves that vary with time as $e^{i\omega t}$ are given by:

$$\omega^2 = l(l+1)\frac{gH}{a^2}. \tag{7.6}$$

Even though there is no Ω in the non-rotating problem (7.4) it is still possible to divide both sides of (7.6) by the (unspecified) term $4\Omega^2$ whose meaning becomes clear only when the frequencies are compared to those on a rotating sphere. Dividing (7.6) by $4\Omega^2$ yields the non-dimensional dispersion relations of non-rotating gravity waves:

$$\omega^2 = \alpha(n+k)(n+k+1), \tag{7.7}$$

in which the non-dimensional frequency equals the dimensional frequency of (7.6) scaled on, the soon to be specified, 2Ω and $l = n+k$.

The exact dispersion relation on a non-rotating sphere (7.7) can be compared to that derived in (6.18) for Inertia-Gravity waves on a rotating sphere $\omega^2 = \alpha(n + 2\delta)^2$ where $2\delta = \frac{1}{2} + (k^2 + \frac{1}{4})^{1/2}$, which was derived in (6.12) for $\alpha n^2 > 1$. For $k \gg \frac{1}{2}$, 2δ can be approximated by $k + \frac{1}{2}$, which yields the approximate dispersion relation:

$$\omega^2 \approx \alpha(n+k+1/2)^2. \tag{7.8}$$

Since $(n+k)(n+k+1) = (n+k+\frac{1}{2})^2 - \frac{1}{4}$ for sufficiently large $n+k$, the non-rotating frequencies in (7.7) are very close to the rotating ones in (7.8) and for $n+k = 5$, the value of $\omega/\alpha^{1/2}$ calculated with (7.7) differ by less than 1 % from that calculated with (7.8): $30^{1/2} \approx 5.48$ versus 5.50. The near-exact match between the exact dispersion relation of gravity waves on a non-rotating sphere, (7.7), and the approximate relation for Inertia-Gravity waves on a rotating sphere, (7.8), at large $n+k$ and for $n^2\alpha > 1$ will be applied to determine the nature of the numerically calculated modes.

Before turning to the identification of the numerically calculated modes it is worth noting that the LSWE is a third-order system in time and therefore the two frequencies of (7.7) are supplemented by the third, $\omega = 0$, eigenvalue. This third $\omega = 0$ frequency is not the trivial solution (i.e., the zero solution where both V and η vanish). Instead, it describes a non-divergent steady zonal velocity, such as $V = (u(\phi), 0)$, balanced geostrophically by the non-vanishing steady η eigenfunction.

Figure 7.3 shows a (slight) zoom-in on the dispersion curves shown in Fig. 7.2 along with the analytic expressions of (7.8) for $(n, k) = (5, 5)$ and $(n, k) = (7, 7)$. Both rotating and non-rotating relations (7.7) (or (6.18)) and (7.8) fall within each of the respective marker and the analytic estimates differ from the numerically derived values by 2–3 % only. Clearly, if the lowest mode in Fig. 7.2 is a Kelvin wave then the meridional mode number of the Inertia-Gravity modes above it should be decreased by 1, so the curves next to the analytic markers should be $n = 4$, instead of $n = 5$, in the case of the datum at $k = 5$ and $n = 6$, instead of $n = 7$, in

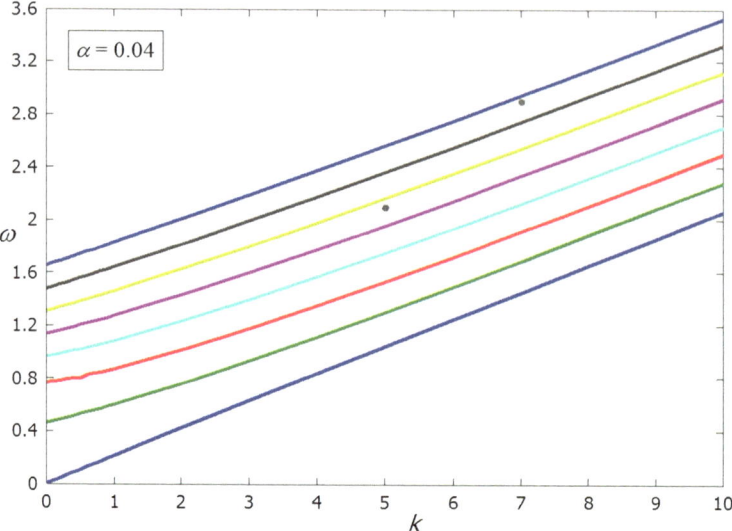

Fig. 7.3 An enlarged version of Fig. 7.2 along with the analytic estimates (7.7) and (7.8) for $\alpha = 0.04$ and $(n, k) = (5, 5)$ and $(7, 7)$. In accordance with the increase in accuracy of (7.8) with k and n, the relative error of the analytic estimate decreases from 3 % at $(n, k) = (5, 5)$ to 1.6 % at $(n, k) = (7, 7)$

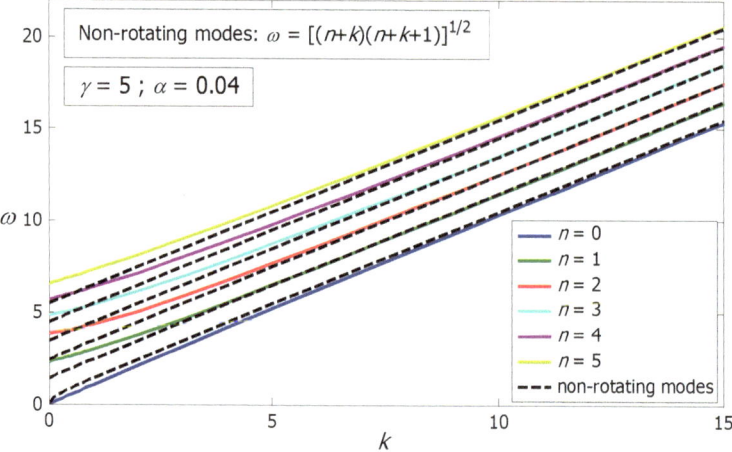

Fig. 7.4 The dispersion relations of Gravity waves on a non-rotating sphere and of Inertia-Gravity waves on a rotating sphere. As anticipated theoretically, the two types of waves differ significantly from one another only at low k and n. $\gamma = \alpha^{-\frac{1}{2}}$ is the square root of Lamb number

the case of the datum at $k = 7$. In both points, decreasing n by 1 to enable the lowest mode in Fig. 7.3 to be a "Kelvin" mode implies a much larger error of the analytic estimates than the 2–3 % error encountered with the original values of n. It should

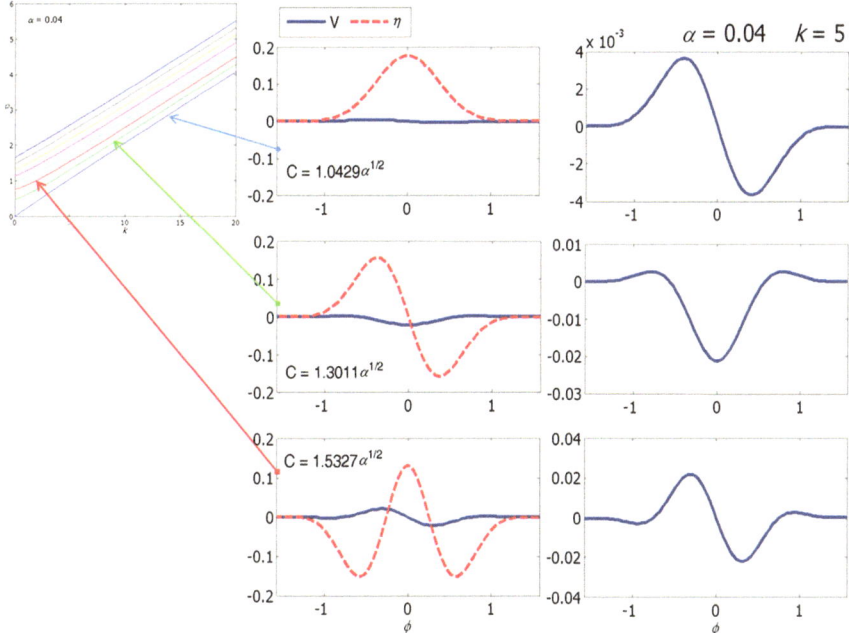

Fig. 7.5 The eigenfunctions corresponding to the lowest three modes of Fig. 7.2 at $k = 5$ and the relation between a pair of eigenfunctions and the associated mode (*inset*). The number of $\eta(\phi)$ zero-crossings increases by 1 when the mode number increases by 1, starting from no zero-crossings in the lowest mode. This count implies that the lowest mode is the $n = 0$ mode of Inertia-Gravity waves. *Note V* does not vanish identically

also be noted that the slope of the lowest $\omega(k)$ curve is larger by 4 % compared to the expected phase speed of Kelvin waves, $\omega = k\sqrt{\alpha}$, which is evident upon examining the intersection of this curve with the right ordinate located at $k = 10$ that occurs at $\omega = 2.08$ instead of $\omega = 2.00$ as expected for Kelvin waves with $\sqrt{\alpha} = 0.2$.

The exact dispersion relations of Gravity waves on a non-rotating sphere, (7.7), are compared in Fig. 7.4 to the numerically calculated relations of Inertia-Gravity waves on a rotating sphere (see Fig. 7.2). It is evident from this comparison that rotation affects the dispersion relation of the first 6 modes displayed in this figure only at low n and k, which is precisely the analytic conclusion derived above based on the close agreement between the exact relation on a non-rotating sphere (7.7) and the approximate relation on a rotating sphere, (7.8), at large k and n. In contrast to the similarity between these dispersion curves of eastward propagating waves on a rotating sphere and those on a non-rotating sphere the dispersion relation on a rotating plane is fundamentally different in that all Inertia-Gravity modes are asymptotic at large k to $\omega = k\sqrt{\alpha}$—the dispersion curve of the eastward propagating Kelvin wave (see Fig. 1.2 and Eq. (1.5)). This comparison implies that eastward propagating waves on a rotating sphere are close to those on a non-rotating sphere more than to those on a rotating plane.

An examination of the eigenfunctions associated with the frequencies calculated in Fig. 7.2 also supports the interpretation of the mode number of Inertia-Gravity waves, n, based on the analytic expressions for the dispersion relations. The number of zero-crossings of $\eta(\phi)$ in Fig. 7.5 clearly equals n. Also, not only is $V(\phi)$ different from zero but the number of its zero-crossings is the same as that of $\partial\eta/\partial\phi$ (i.e., larger by 1 compared to the number of zero-crossings of $\eta(\phi)$), which is consistent with the balance of terms in the η (i.e., second) equation of (4.4) where for large C, the contribution of η to V is small compared to that of $\partial\eta/\partial\phi$ (i.e., $\tan\phi/C < 1$).

The foregoing numerically determined association between the number of zero-crossings of η and the mode number in Fig. 7.2 as well as the accuracy of the analytic expressions (that, as expected, increases with k) indicates that the lowest mode in Fig. 7.2 is the $n = 0$ Inertia-Gravity mode and not a "Kelvin" wave (and certainly not a Kelvin wave).

Reference

Boyd JP, Zhou C (2008) Uniform asymptotics for the linear Kelvin wave in spherical geometry. J Atmos Sci 65:655–660

Index

A
Airy equation
    functions, 20–22, 48

B
β-plane—equatorial
    midlatitude, 1–3, 5, 7, 8, 11, 12, 16, 25, 27,
    29, 34, 47, 53, 69

C
Coriolis frequency, 1, 2, 4, 9, 11, 19, 35, 45, 47

F
f-plane, 1, 2, 5, 6, 11, 22

H
Harmonic waves, 13, 22, 27, 29, 34, 49, 51, 53, 56
Hermite functions, 1, 7, 8, 29–31, 62
Hydrostatic balance, 2

G
Geostrophic balance, 5
Gravity waves, 4, 13, 31, 58, 72–75

I
Inertia-gravity waves, 4–8, 10, 11, 18, 19, 23,
    24, 32, 33, 42, 57, 59, 61, 64, 66, 67,
    69, 71–76
Inertial waves, 19

K
Kelvin waves, 3, 5, 7, 8, 11, 14, 15, 33, 34, 37,
    45, 69–72, 75, 76

L
Laplace tidal equation, 1

M
Mixed mode (on the equator), 7, 8, 34, 43–45

P
Planetary waves, 5–7, 10, 11, 15, 18, 19, 23,
    25–27, 33, 34, 38, 42, 57, 59, 61, 64,
    65, 67
Poincarè waves, 4, 6, 24, 38, 39, 66

R
Rossby waves, 5, 6, 8, 17, 25, 39, 49, 51

S
Shallow water equations, 1, 10, 11, 35, 55, 70
Spherical harmonic functions, 72

T
Trapped waves, 13, 19, 22–25, 27, 29, 48, 49,
    51–53, 56

© The Author(s) 2015
N. Paldor, *Shallow Water Waves on the Rotating Earth*,
SpringerBriefs in Earth System Sciences, DOI 10.1007/978-3-319-20261-7